The Handbook of Environmental Chemistry

Volume 3 Anthropogenic Compounds Part L

O. Hutzinger
Editor-in-Chief

Springer-Verlag Berlin Heidelberg GmbH

Endocrine Disruptors
Part I

With contributions by
H. G. Dörr · J. Dötsch · K. W. Gaido · C. A. Harris
W. R. Kelce · K. S. Korach · C. M. Markey
D. P. McDonnell · M. Metzler · C. L. Michaelson
S. O. Mueller · E. Pfeiffer · S. Safe · C. Sonnenschein
A. M. Soto · J. P. Sumpter · L. U. Thompson
W. E. Ward · L. Wildt · E. M. Wilson

Springer

Environmental chemistry is a rather young and interdisciplinary field of science. Its aim is a complete description of the environment and of transformations occurring on a local or global scale. Environmental chemistry also gives an account of the impact of man's activities on the natural environment by describing observed changes.

"The Handbook of Environmental Chemistry" provides the compilation of today's knowledge. Contributions are written by leading experts with practical experience in their fields. The Handbook will grow with the increase in our scientific understanding and should provide a valuable source not only for scientists, but also for environmental managers and decision makers.

As a rule, contributions are specially commissioned. The editors and publishers will, however always be pleased to receive suggestions and supplements information. Papers for *The Handbook of Environmental Chemistry* are accepted in English.

In reference The Handbook of Environmental Chemistry is abbreviated Handb. Environ. Chem. and is cited as a journal.

Springer WWW home page: http://www.springer.de
Visit the MS home page at http://link.springer.de/series/hec/
or http://link.springer.ny.com/series/hec/

ISSN 1433-6847
ISBN 978-3-662-14681-1

Library of Congress Cataloging-in-Publication Data
The Natural environment and the biogeochemical cycles / with
contributions by P. Craig ... [et al.].
v. <A–F > : ill. ; 25 cm. – (The Handbook of environmental chemistry :
v. 1) Includes bibliographical references and indexes.
ISBN 978-3-662-14681-1   ISBN 978-3-540-48428-8 (eBook)
DOI 10.1007/978-3-540-48428-8
1. Biogeochemical cycles. 2. Environmental chemistry.
I. Craig, P. J., 1944– . II. Series.
QD31. H335 vol. 1   [QH344]   628.5 s

Production Editor: Christiane Messerschmidt, Rheinau
Cover Design: E. Kirchner, Springer-Verlag
Typesetting: Fotosatz-Service Köhler GmbH, Würzburg
Printed on acid-free paper   SPIN: 10690734   52/3020 – 5 4 3 2 1 0

The Handbook of Environmental Chemistry Now Also Available Electronically

For all customers with a standing order for The Handbook of Environmental Chemistry we offer the electronic form via LINK free of charge. Please contact your librarian who can receive a password for free access to the full articles. By registration at:

http://link.springer.de/series/hec/reg_form.htm

However, if you do not have a standing order, you can browse through the table of contents of the volumes and the abstracts of each article at:

http://link.springer.de/series/hec/
http://link.springer-ny.com/series/hec/

There you will also find information about the

- Editorial Bord
- Aims and Scope
- Instructions for Authors

Preface

Environmental Chemistry is a relatively young science. Interest in this subject, however, is growing very rapidly and, although no agreement has been reached as yet about the exact content and limits of this interdisciplinary discipline, there appears to be increasing interest in seeing environmental topics which are based on chemistry embodied in this subject. One of the first objectives of Environmental Chemistry must be the study of the environment and of natural chemical processes which occur in the environment. A major purpose of this series on Environmental Chemistry, therefore, is to present a reasonably uniform view of various aspects of the chemistry of the environment and chemical reactions occurring in the environment.

The industrial activities of man have given a new dimension to Environmental Chemistry. We have now synthesized and described over five million chemical compounds and chemical industry produces about hundred and fifty million tons of synthetic chemicals annually. We ship billions of tons of oil per year and through mining operations and other geophysical modifications, large quantities of inorganic and organic materials are released from their natural deposits. Cities and metropolitan areas of up to 15 million inhabitants produce large quantities of waste in relatively small and confined areas. Much of the chemical products and waste products of modern society are released into the environment either during production, storage, transport, use or ultimate disposal. These released materials participate in natural cycles and reactions and frequently lead to interference and disturbance of natural systems.

Environmental Chemistry is concerned with reactions in the environment. It is about distribution and equilibria between environmental compartments. It is about reactions, pathways, thermodynamics and kinetics. An important purpose of this Handbook, is to aid understanding of the basic distribution and chemical reaction processes which occur in the environment.

Laws regulating toxic substances in various countries are designed to assess and control risk of chemicals to man and his environment. Science can contribute in two areas to this assessment; firstly in the area of toxicology and secondly in the area of chemical exposure. The available concentration ("environmental exposure concentration") depends on the fate of chemical compounds in the environment and thus their distribution and reaction behaviour in the environment. One very important contribution of Environmental Chemistry to the above mentioned toxic substances laws is to develop laboratory test methods, or mathematical correlations and models that predict the environmental fate of new chemical compounds. The third purpose of this Handbook is to help in the basic understanding and development of such test methods and models.

The last explicit purpose of the Handbook is to present, in concise form, the most important properties relating to environmental chemistry and hazard assessment for the most important series of chemical compounds.

At the moment three volumes of the Handbook are planned. Volume 1 deals with the natural environment and the biogeochemical cycles therein, including some background information such as energetics and ecology. Volume 2 is concerned with reactions and processes in the environment and deals with physical factors such as transport and adsorption, and chemical, photochemical and biochemical reactions in the environment, as well as some aspects of pharmacokinetics and metabolism within organisms. Volume 3 deals with anthropogenic compounds, their chemical backgrounds, production methods and information about their use, their environmental behaviour, analytical methodology and some important aspects of their toxic effects. The material for volume 1, 2 and 3 was each more than could easily be fitted into a single volume, and for this reason, as well as for the purpose of rapid publication of available manuscripts, all three volumes were divided in the parts A and B. Part A of all three volumes is now being published and the second part of each of these volumes should appear about six months thereafter. Publisher and editor hope to keep materials of the volumes one to three up to date and to extend coverage in the subject areas by publishing further parts in the future. Plans also exist for volumes dealing with different subject matter such as analysis, chemical technology and toxicology, and readers are encouraged to offer suggestions and advice as to future editions of "The Handbook of Environmental Chemistry".

Most chapters in the Handbook are written to a fairly advanced level and should be of interest to the graduate student and practising scientist. I also hope that the subject matter treated will be of interest to people outside chemistry and to scientists in industry as well as government and regulatory bodies. It would be very satisfying for me to see the books used as a basis for developing graduate courses in Environmental Chemistry.

Due to the breadth of the subject matter, it was not easy to edit this Handbook. Specialists had to be found in quite different areas of science who were willing to contribute a chapter within the prescribed schedule. It is with great satisfaction that I thank all 52 authors from 8 countries for their understanding and for devoting their time to this effort. Special thanks are due to Dr. F. Boschke of Springer for his advice and discussions throughout all stages of preparation of the Handbook. Mrs. A. Heinrich of Springer has significantly contributed to the technical development of the book through her conscientious and efficient work. Finally I like to thank my family, students and colleagues for being so patient with me during several critical phases of preparation for the Handbook, and to some colleagues and the secretaries for technical help.

I consider it a privilege to see my chosen subject grow. My interest in Environmental Chemistry dates back to my early college days in Vienna. I received significant impulses during my postdoctoral period at the University of California and my interest slowly developed during my time with the National Research

Council of Canada, before I could devote my full time of Environmental Chemistry, here in Amsterdam. I hope this Handbook may help deepen the interest of other scientists in this subject.

Amsterdam, May 1980 *O. Hutzinger*

Twentyone years have now passed since the appearance of the first volumes of the Handbook. Although the basic concept has remained the same changes and adjustments were necessary.

Some years ago publishers and editors agreed to expand the Handbook by two new open-end volume series: Air Pollution and Water Pollution. These broad topics could not be fitted easily into the headings of the first three volumes. All five volume series are integrated through the choice of topics and by a system of cross referencing.

The outline of the Handbook is thus as follows:

1. The Natural Environment and the Biochemical Cycles,
2. Reaction and Processes,
3. Anthropogenic Compounds,
4. Air Pollution,
5. Water Pollution.

Rapid developments in Environmental Chemistry and the increasing breadth of the subject matter covered made it necessary to establish volume-editors. Each subject is now supervised by specialists in their respective fields.

A recent development is the accessibility of all new volumes of the Handbook from 1990 onwards, available via the Springer Homepage http://www.springer. de or http://Link.springer.de/series/hec/ or http://Link.springerny.com/ series/hec/.

During the last 5 to 10 years there was a growing tendency to include subject matters of societal relevance into a broad view of Environmental Chemistry. Topics include LCA (Life Cycle Analysis), Environmental Management, Sustainable Development and others. Whilst these topics are of great importance for the development and acceptance of Environmental Chemistry Publishers and Editors have decided to keep the Handbook essentially a source of information on "hard sciences".

With books in press and in preparation we have now well over 40 volumes available. Authors, volume-editors and editor-in-chief are rewarded by the broad acceptance of the "Handbook" in the scientific community.

Bayreuth, July 2001 *Otto Hutzinger*

Contents

Foreword XIII

Mechanisms of Estrogen Receptor-Mediated Agonistic and Antagonistic Effects
S. O. Mueller, K. S. Korach 1

In Vitro Methods for Characterizing Chemical Interactions with Steroid Hormone Receptors
K. W. Gaido, D. P. McDonnell, S. Safe 27

Antiandrogenic Effects of Environmental Endocrine Disruptors
W. R. Kelce, E. M. Wilson 39

Chemistry of Natural and Anthropogenic Endocrine Active Compounds
M. Metzler, E. Pfeiffer 63

Exposure to Endogenous Estrogens During Lifetime
J. Dötsch, H. G. Dörr, L. Wildt 81

Dietary Estrogens of Plant and Fungal Origin: Occurrence and Exposure
W. E. Ward, L. U. Thompson 101

Alkylphenols and Bisphenol A as Environmental Estrogens
C. M. Markey, C. L. Michaelson, C. Sonnenschein, A. M. Soto 129

Hydroxylated Polychlorinated Biphenyls (PCBs) and Organochlorine Pesticides as Potential Endocrine Disruptors
S. Safe 155

The Endocrine Disrupting Potential of Phthalates
C. A. Harris, J. P. Sumpter 169

Subject Index 203

Contents of Part II

Estrogen and Human Breast Cancer
Y. F. Hu, I. H. Russo, J. Russo

The Role of Sex Hormones in Prostate Cancer
M. C. Bosland

Beneficial and Adverse Effects of Dietary Estrogens on the Human Endocrine System: Clinical and Epidemiological Data
D. M. Tham

Mechanism-Based Carcinogenic Risk Assessment of Estrogens and Estrogen-Like Compounds
B. A. T. Willems, C. J. Portier, G. W. Lucier

Alterations in Male Reproductive Development: The Role of Endocrine Disrupting Chemicals
S. H. Swan, F. vom Saal

Effects of Perinatal Estrogen Exposure on Fertility and Cancer in Mice
R. R. Newbold

Genotoxic Potential of Natural and Synthetic Endocrine Active Compounds
M. Metzler

Emerging Issues Related to Endocrine Disrupting Chemicals and Environmental Androgens and Antiandrogens
L. E. Gray Jr., C. Lambright, L. Parks, R. W. Tyl, E. F. Orlando, L. J. Guilette Jr., C. J. Wolf, J. C. Seely, T.-Y. Chang, V. Wilson, A. Hotchkiss, J. S. Ostby

Developmental and Reproductive Abnormalities Associated with Endocrine Disruptors in Wildlife
E. F. Orlando, L. J. Guillette Jr.

Endocrine Disruption in the Aquatic Environment
J. P. Sumpter

Foreword

Endocrine disruptors, also called endocrine-active compounds, endocrine modulators, environmental hormones, hormone-related toxicants etc., are compounds that exhibit the potential to interfere with the endocrine system of humans and animals. The endocrine system uses endogenous hormones, i.e. compounds produced by certain glands, to communicate with various tissues and regulate body functions such as growth, development and reproduction. The group of endocrine disruptors comprises a large and still increasing number of natural and anthropogenic agents with diverse chemical structures.

Wide scientific and public interest in endocrine disruptors has evolved about ten years ago, when evidence that chemicals may adversely affect the sexual development of a number of wildlife species was presented at a Workshop convened by Theo Colburn (Proceedings: T. Colburn and C.R. Clement, eds. *Chemically-induced Alterations in Sexual and Functional Development: The Wildlife/Human Connection.* Princeton, NJ: Princeton Scientific Publishing Co. 1992). Subsequent reports, showing that numerous everyday chemicals exhibit hormonal activity, and linking male reproductive problems such as low sperm counts and increased rates of testicular cancer in young men to environmental hormones, increased the concern about endocrine disruptors. As a consequence, endocrine disruptors have become over the past years, and will continue to be over the next years, a "hot" topic at toxicological Meetings, in public media, and also in the political arena.

Although the general interest in endocrine disruptors is relatively recent, scientific interest dates back to the sixties and early seventies when the adverse effects of a synthetic estrogen, diethylstilbestrol (DES), on experimental animals and also on humans were reported for the first time. Based on these observations, a Confererence was convened by John A. McLachlan in 1979, at which the basic questions of endocrine disruptors, i.e. "what an estrogen is and how it works, and what effects estrogenic substances might have on human health" (cited from the foreword of the Proceedings: John A. McLachlan, ed. *Estrogens in the Environment.* New York: Elsevier North Holland Inc. 1980) were already raised and addressed for estrogenic compounds, which still constitute the major group of endocrine disruptors. The environmental occurrence and impact of estrogenic agents of natural and man-made origin were also extensively discussed at this Meeting and at a subsequent Conference in 1985 (Proceedings: John A. McLachlan, ed. *Estrogens in the Environment II. Influences on Development.* New York: Elsevier Science Publishing Co., Inc. 1985). These two Conferences and the pioneering work of John McLachlan, Howard Bern and a few other investigators have laid the ground for the present endocrine disruptor field.

Research in this area so far has clearly shown that the answers for the two basic questions of endocrine disruptors, posed above for estrogens, will not be easy and straightforward. The structural requirements and mechanisms of action of endocrine disruptors are complicated by the fact that multiple ways exist to interfere with the endocrine system. For example, endocrine disruptors can (i) mimic or block the binding of endogenous hormones to their receptors, (ii) affect cell signaling pathways in a direct, i.e. non-receptor-mediated manner, or (iii) alter the production or metabolism of endogenous hormones. The effects on human and animal health may be even more difficult to assess due to the complexity of the endocrine system. They appear to depend on the chemical structure and dose of the individual compound, the duration of exposure, and the species, developmental stage (age) and gender of the organism. The exposure to hormonally active compounds does not necessarily lead to adverse effects, as is demonstrated by the putative anti-carcinogenic effects of certain plant estrogens in Asian populations. To sort out the adverse and beneficial effects of endocrine disruptors and the underlying mechanisms will be a challenging scientific and also an important practical task, since exposure to such compounds through food, air, water and many household products is ubiquitous and unavoidable.

The present book provides an overview on important aspects of endocrine disruptors. The handbook comprises 19 chapters and is divided into two parts. Part I addresses the mechanisms and detection of hormone action, and the chemistry of and exposure to the various classes of natural and anthropogenic endocrine disruptors. Part II focuses on the association of sex hormones with diseases in humans, on the effects of endocrine-active compounds in experimental systems, and on the association of endocrine disruptors with environmental effects.

The chapters were authored by scientists who are highly recognized in their areas of research. The editor is greatly indebted to all of them as well as to the staff of Springer for their commitment. It is hoped that the result of this joint effort may prove as an useful and inspiring source of information for everybody interested in the multifaceted issue of endocrine disruptors.

Karlsruhe, June 2001 Manfred Metzler

CHAPTER 1

Mechanisms of Estrogen Receptor-Mediated Agonistic and Antagonistic Effects

Stefan O. Mueller, Kenneth S. Korach

National Institutes of Health, National Institute of Environmental Health Sciences, Laboratory of Reproductive and Developmental Toxicology-Receptor Biology Group, PO Box 12233, Research Triangle Park, NC 27709, USA
e-mail: korach@niehs.nih.gov

Estrogenic compounds exert a vast variety of effects in wildlife and humans. Endogenous estrogens, like estradiol, regulate growth and development of their target tissues. Exogenous compounds with estrogenic and/or anti-estrogenic activities may disrupt these regulatory pathways. Environmental or industrial chemicals and phytoestrogens interfering with the hormonal or endocrine system are defined as *endocrine disruptors*. The estrogen receptor is the major regulatory unit within the estrogen-signaling pathway. Effects mediated by the estrogen receptor are not solely defined by its ligand but are rather modulated by the tissue expression of the receptor, and the cellular and genetic environment. *Endocrine disruptors* are able to mimic estrogens by binding to the estrogen receptor and induce or inhibit estrogenic response. Phenolic compounds, pesticides, phytoestrogens, and synthetic estrogens like diethylstilbestrol are examples of *endocrine disruptors* or so-called *endocrine active compounds*. The potential of these chemicals to interfere with the endocrine system is primarily defined by their interaction with the estrogen receptor. However, the observed physiological effects are the result of a complex interplay between the estrogen receptor, its ligand, co-regulators, and other cell signaling pathways dependent on the target tissue.

Keywords. SERM, Endocrine active compounds, Estrogens, Anti-estrogens, Xenoestrogens, Tissue-specificity

1 **Introduction** . . . 3

2 **Nuclear Receptor Superfamily** . . . 3

3 **Estrogen Receptors *α* and *β*** . . . 4

4 **Tissue Distribution of ER*α* and ER*β*** . . . 5

5 **Mechanism of Estrogen Receptor-Mediated Hormone Action** . . . 7

5.1 Binding of Ligand and Ligand-Dependent AF-2 . . . 8
5.2 Dimerization and DNA Binding of the ER . . . 9
5.3 Transcriptional Activation Via AF-1 and AF-2 . . . 10
5.4 Co-Activators and -Repressors . . . 10
5.5 Estrogen Responsive Elements (EREs) . . . 11
5.6 Cross-Talk with Other Signaling Pathways . . . 11
5.7 Estrogen-Regulated Genes . . . 11

The Handbook of Environmental Chemistry Vol. 3, Part L
Endocrine Disruptors, Part I
(ed. by M. Metzler)

6 **Substances Interacting with the ER** 12

7 **Mechanisms of Receptor-Agonists and/or -Antagonists** 16

8 **Endocrine Active Compounds: Estrogens or Antiestrogens, or Both?** 20

9 **References** 21

List of Abbreviations

AF	activation function
AP-1	activator protein 1
cERE	consensus estrogen responsive element
CoA	co-activator
DES	diethylstilbestrol
DBD	DNA binding domain
E	estrogen
EAC	endocrine active compound
EGF	epidermal growth factor
ER	estrogen receptor
ERE	estrogen responsive element
ERKO	estrogen receptor knock-out mouse
GF	growth factor
GTA	general transcription assembly
Hsp	heat-shock protein
IGF	insulin-like growth factor
LBD	ligand binding domain
MAPK	mitogen-activated protein kinase
nERE	non-consensus estrogen responsive element
NLS	nuclear localization signal
NR	nuclear receptor
OH-PCB	polychlorinated hydroxy-biphenyls
OHT	4-hydroxytamoxifen
PK	protein kinase
Pol	polymerase
RAP	receptor associated protein
RRE	raloxifene response element
SERM	selective estrogen receptor modulator
SHR	steroid hormone receptor
TBP	TATA binding protein
TeBG	testosterone binding globulin
TF	transcription factor

1 Introduction

Estrogens exert a vast variety of effects in wildlife and humans. They regulate development and growth by inducing cell proliferation and cell differentiation. Observations of developmental abnormalities in wildlife exposed to chemicals and rising incidences of hormone-dependent cancers in humans have raised concerns in the public and the scientific community [1, 2]. Experimental data on chemicals like diethylstilbestrol (DES) [3] confirmed the potential of xenobiotics to impair the endocrine system, especially estrogen function, and to cause developmental abnormalities and cancer.

How is hormonal regulation disrupted by exogenous compounds? The endocrine system utilizes hormones secreted by the specific hormonal glands directly into the bloodstream. The secreted hormones are in part bound to serum proteins and transported to the target organ or cell [4]. These hormones can also be metabolized, resulting in either an activation or deactivation. The response of the cell to the active hormone is dependent for the most part on the presence of specific hormone receptors. The affinity and specificity of the respective receptor to the external signal (hormone) determines whether the particular hormone is bound by the receptor. The receptor with its bound ligand, i.e., the hormone, transmits the signal resulting in a specific biological response. In the case of steroid hormones the main cellular response is an activation of gene transcription (genomic effects), although "non-genomic," rapidly occurring effects have also been discussed [5]. Any exogenous substance could potentially interfere with any of the steps described above to cause a disruption of the endocrine system. These substances are therefore referred to as *endocrine disruptors.*

In this chapter we will focus on the last step of the cascade of hormone action, namely receptor mediated effects of xenobiotics. Xenobiotics can mimic steroids by binding to the receptor and induce (agonize) or inhibit (antagonize) the steroid response. These different actions are not only due to the respective compound but are dependent on the cellular and genetic context and, probably, on the receptor subtype present in the target cell or tissue.

2 Nuclear Receptor Superfamily

Hormone receptors comprise membrane-bound receptors, like trophic peptide hormone (e.g., gonadotropin, insulin, and growth hormone) and growth factor (e.g., epidermal growth factor (EGF) and insulin-like GF-I (IGF-I)) receptor, and nuclear receptors. Nuclear receptors are located within the nucleus of the cell. The nuclear receptor superfamily consists of steroid hormone, thyroid, retinoid, and a growing number of orphan receptors with unknown physiological ligands [6–8].

The steroid hormone receptors (SHR) comprise the glucocorticoid, mineralcorticoid, progesterone, androgen, and estrogen receptors [8]. Steroid hormones play a major part in the regulation of development, homeostasis, and

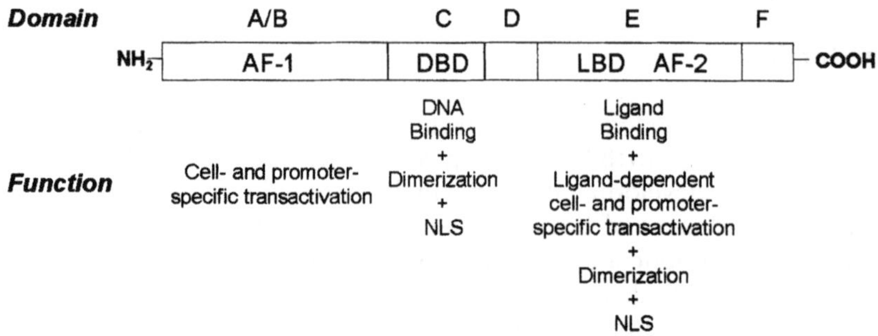

Fig. 1. Modular structure of steroid hormone receptors. Distinct domains are designated A–F. The NH_2-terminal domain A/B contains the ligand-independent activation function (AF-1); the highly conserved DNA binding domain (DBD) recognizes specific DNA sequences (hormone response elements) and is located in domain C, which also contains a dimerization and nuclear localization signal (NLS). Domain E is connected by the so-called *hinge* region (D) and is responsible for ligand binding, dimerization, and also contains a ligand-dependent NLS. The ligand-dependent activation function (AF-2) is also located within domain E. The F domain is highly variable between different species

stimulate growth and differentiation of their target cells [4]. These effects are mediated by the respective SHR. The SHRs, like other members of the nuclear receptor superfamily, have a modular structure, i.e., they are composed of distinct functional domains (Fig. 1) [8–10]. Functionally, SHRs can be characterized as ligand-inducible transcription factors. Upon binding of the steroid hormone (ligand) to its cognate receptor (LBD, domain E, Fig. 1), the receptor dimerizes (domains C and E) and binds to specific hormone responsive DNA sequences (DBD, domain C, Fig. 1). The DNA bound SHR activates in concert with co-regulators and, depending on the cell type and the promoter, specific genes, which are consequently defined as hormone-regulated genes.

In the following sections we will describe and discuss the effects mediated by the estrogen receptor.

3
Estrogen Receptors α and β

After the cloning of the human estrogen receptor (ER), ERα, in 1986 [11, 12] it was well accepted that there was only one form of the ER. Surprisingly, a second ER, termed ERβ, was discovered [13], which was identified in cDNA libraries of rat prostate [13] and human testis [14]. In Fig. 2 the protein structure and homology between the estrogen receptors from human and mouse are given.

ERα and ERβ show high sequence homologies in the DNA binding domain (>90%) and to a lesser extent in the ligand binding domain (~60%). The human ERα (hERα) is a 595-amino acid-containing protein, with high interspecies homologies (88% to the mouse ERα, Fig. 2) [15]. The originally cloned human ERβ [14] represents a 477-amino acid-protein. Shortly after that the full-length cDNA for hERβ was cloned, consisting of 485 amino acids [16], and

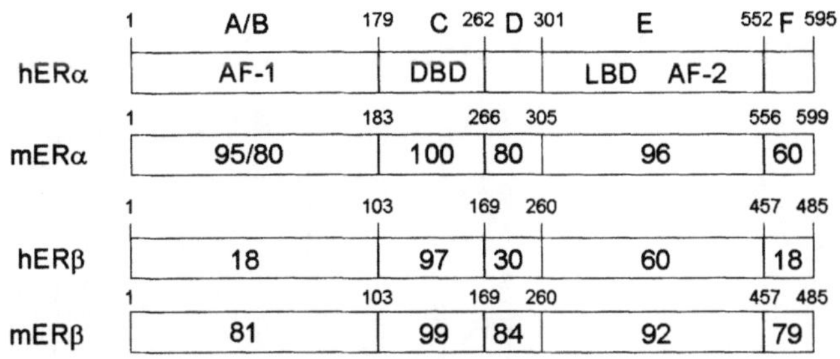

Fig. 2. Homologies between human (h) and mouse (m) ERα and ERβ. The modular domain structure of the ER is shown schematically on the *top* (see Fig. 1). Homologies are given in percent for each domain for hERα. mERα and hERβ are compared to hERα, and mERβ is compared to hERβ. Numbers indicate the numbers of amino acids at borders between domains

now appears to be the "short" form of the ERβ. The "long" form contains 45 additional amino acids at the amino terminus belonging to AF-1 (domain A, Fig. 1) [17–19]. Physiologically, the "long" form is considered possibly to be the predominant ERβ. However, in vitro data of different laboratories indicate that no striking differences in their biological activity exist.

Normal and neoplastic estrogen target tissues not only express ERα and/or ERβ, but also express ER mutants and variants. In different neoplastic tissues a vast variety of receptor mutants are found [20–27]. These mutants include receptors with point mutations, truncated transcripts, or insertional mutations. Data indicate that some mutants are stably expressed in vivo, which would support the idea that ER mutants are functional in the estrogen-signaling pathway. A mutated receptor, which shows increased activity without any ligand (constitutively active receptor) could be responsible for estrogen-independent growth of breast tumors. Indeed, in estrogen-unresponsive breast cancer specimens a truncated ERα was found to be highly expressed [24]. On the other hand, mutations in the ER could lead to an increased ligand-dependent activity. Catherino and colleagues [28] showed that a naturally occurring mutant exhibited increased activity in breast cancer cells when treated with an antiestrogen. These results indicate that receptor variants may play a highly influential role in tumor development and progression.

4 Tissue Distribution of ERα and ERβ

Since estrogenic action is mediated mainly through the ER (see following section) an estrogen target tissue is defined as one expressing functional estrogen receptor and showing a distinct response when exposed to estrogen. The tissue distribution of the ER also characterizes the potential target for any compound, agonist and/or antagonist, acting through the ER. In Fig. 3, a qualitative summary of the tissue distribution and abundance of ERα and ERβ is given. The data shown were compiled from humans, mice, and rats, and are given representatively for humans. Figure 3 is not intended to give a precise quantitative

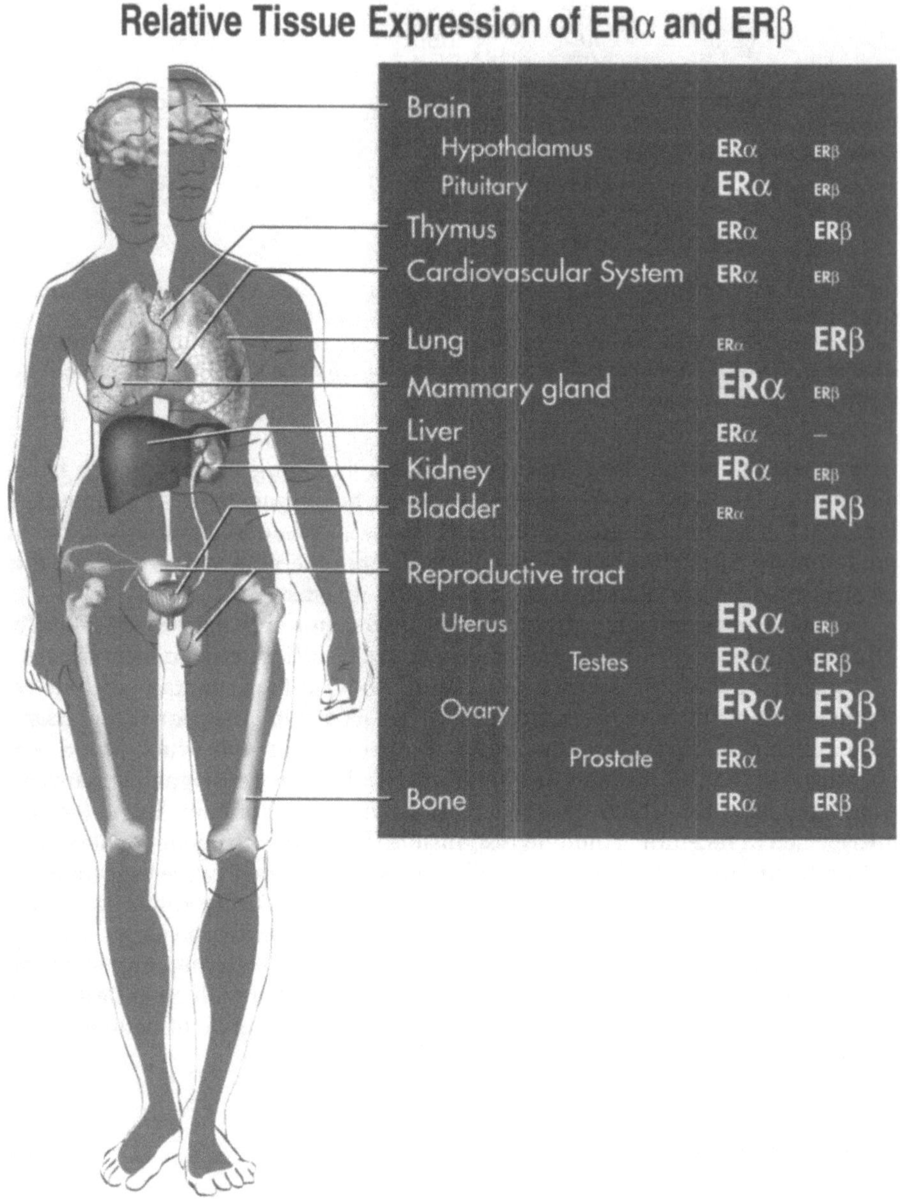

Fig. 3. Examples of qualitative tissue expression of ERα and ERβ. The font sizes of the respective ER reflect the relative expression of ERα and ERβ in each tissue. The figure summarizes published data on cells and tissues of humans, mice, and rats. Data include RNA- and protein expression (data from [14, 29–40]

measure, but rather should be interpreted as an estimate overview for ER expression (data from [14, 29–40]).

The diversity of estrogen target tissues is illustrated in Fig. 3. Furthermore, there are certain tissues with predominant expression of ERβ, that are mainly, with the exception of the ovary, non-classical target tissues [40]. The knock-out mice for ERα (αERKO) and ERβ (βERKO) are excellent models to determine the importance of the respective receptor in development, normal physiology, and neoplasia [41–44]. Studies done with these knock-out mice indicate that most aberrant phenotypes – hypotrophy of the uterus, infertility, and rudimentary mammary gland development, to mention only a few – appear to be dependent on ERα. Studies so far showed that the major influence of ERβ in the female is restricted to ovarian function [41, 44]. Further studies on the effect of ERβ (and ERα) in other tissues (e.g., skeletal, immune-, cardiovascular system, and male reproductive tract) will define the role of ERβ in signaling and its effects. The presence of the respective estrogen receptor in a specific tissue leads to a potential target-site for pharmaceutical drugs and xenobiotics.

5 Mechanism of Estrogen Receptor-Mediated Hormone Action

Most endogenous estrogen (17β-estradiol and estrone) in women is produced in the granulosa cells of the ovary. Precursor molecules, like androstenedione, secreted by the adrenal gland, can be metabolized by aromatization to estrone (e.g., in adipose tissue or brain). During pregnancy the placenta produces large amounts of 17β-estradiol (estradiol). The plasma level of estradiol, the most predominant estrogen, is around 20–60 pg/ml in adult, premenopausal women (maximum during the menstrual cycle up to 700 pg/ml), which drops below 20 pg/ml during postmenopause. Males also produce estradiol (~20 pg/ml, young male). In males, estrogen is predominantly formed by aromatization of androstenedione (via estrone) in peripheral tissues with a small amount secreted by the testes. Estrone is metabolized by 17β-hydroxysteroid dehydrogenase to estradiol [45].

Estradiol and estrone are, like exogenous estrogens or antiestrogens, transported via the bloodstream to their target cell. Since estrogens are water insoluble, circulating estrogens are bound to the carrier proteins albumin with low affinity and with higher affinity to testosterone binding globulin (TeBG). Only free estrogen can enter the cell by diffusion due to its apolarity (Fig. 4A). Therefore, not the total plasma level of estrogen, but the free, unbound estrogen level determines the amount of estrogen capable of entering the cell and inducing a specific response. Figure 4 gives a schematic overview of the molecular mechanism of estrogen receptor action.

The capacity of a specific cell to respond to estrogen is dependent on the presence of its specific receptor, ERα or ERβ. Although the ERs show differential tissue distribution and may also exert different effects, described below, the high degree of homology contributes to the general opinion that the molecular mechanism of ER action is conserved.

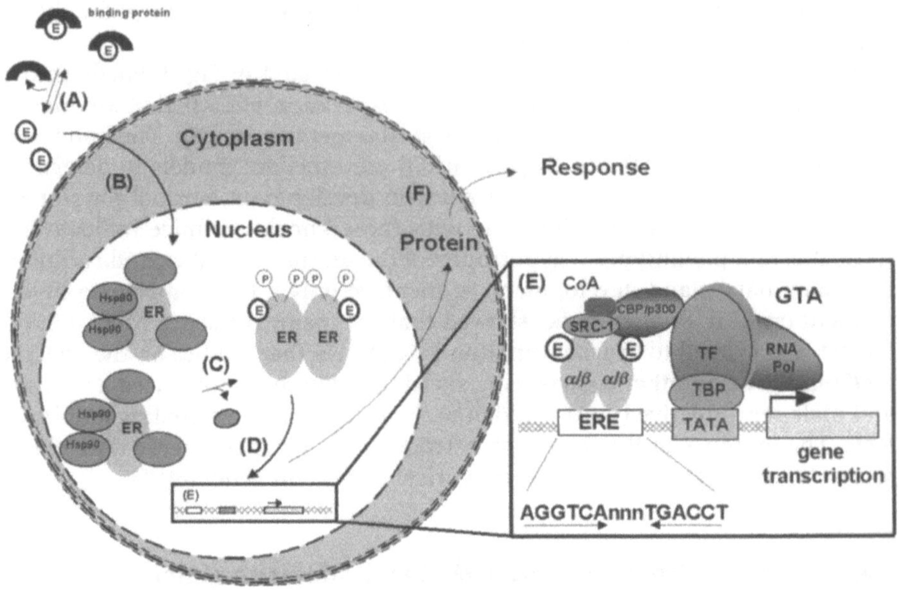

Fig. 4. Mechanism of estrogen receptor action. (A) Estrogens (E) like estradiol are bound to carrier proteins in the serum. Upon dissociation from these proteins, (B) E can pass the cell membrane easily by diffusion due to their apolarity. The estrogen receptor (ER) is located within the nucleus. Inactive ER is bound to various receptor associated proteins (RAP), like heat-shock proteins (Hsp90). (C) The RAPs are displaced and the ligand (E) binds to the ER. The ER dimerizes and (D) the dimer binds with its DNA binding domains to the estrogen responsive element (ERE) located within the control region of the target gene promoter. (E) In concert with the *general transcription assembly* (GTA), which comprises various transcription factors (TF), the RNA polymerase (RNA Pol) and other proteins, specific genes are transcribed into the respective mRNA. The GTA binds through the TATA binding protein (TBP) to the TATA box within the promoter. Co-activators (CoA) such as CBP/p300 and SRC-1 most likely link the ER-dimer with the GTA. (F) Translation of the mRNA results in a specific protein (e.g., growth factors) which results in the cell-specific and promoter-specific response (see text for more details)

5.1 Binding of Ligand and Ligand-Dependent AF-2

The ER is located within the nucleus. In its inactive state the receptor is bound to different chaperones like heat shock proteins (Hsp90) and immunophilins (p59) [46, 47]. It is believed that steroid receptors undergo a so-called *chaperone cycle* resulting in a mature complex. The receptor protein is stabilized in its hormone binding conformation within the mature complex from which the ER is spontaneously released [46]. The free receptor can then bind hormone (Fig. 4C). The ER also undergoes further modifications like phosphorylation [48] (reviewed in [49]). More recent data provided further evidence for the need of phosphorylation for ER-mediated activation [50, 51]. However, a clear importance or requirement for receptor signaling is not yet fully evaluated.

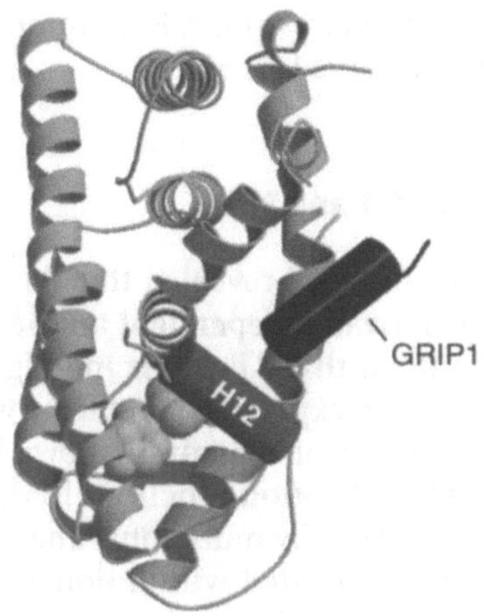

Fig. 5. Overall structure of the DES-ERα LBD-GRIP1 complex. The LBD is shown as ribbon drawing. The co-activator peptide GRIP1 is colored *black* and helix 12 (labeled H12) is colored *gray*, both shown as cylinders. DES is shown in space-filling representation. Drawn according to [53] using Molscript with the coordinates from the Brookhaven protein databank (accession 2ERD)

Estrogen fits into the hydrophobic cavity of the ligand binding domain (LBD) of the ER. Helix 12 of the LBD functions as a "lid" sealing the hydrophobic cleft and stabilizing the binding of estrogen (Fig. 5) [52, 53]. Conformational changes in the protein LBD allow an active, ligand-dependent transactivation function AF-2 to occur [53, 54]. Katzenellenbogen and colleagues showed that the LBD and the AF-2 are structurally distinct units, although they are located within one protein domain (see also Fig. 1) [55].

5.2
Dimerization and DNA Binding of the ER

Dimerization of the ER is dependent on specific regions within the C- and E-domains (Fig. 1). The ER binds through its DNA binding domain (DBD) to a specific DNA sequence, termed an estrogen response element (ERE) (Fig. 4D, E). The consensus ERE (cERE) consists of an inverted repeat of six bases with a three base spacer (Fig. 4E) [8, 9]. This inverted repeat allows the binding of the ER dimer. The DBD contains two zinc finger motifs. In each motif the zinc ion is complexed by four cysteine residues connected by a short β-sheet (between cysteine 1 and 2) and by a α-helical structure (between cysteine 2, 3, and 4). The latter one, the so-called P-box in motif one is responsible for binding and discrimination of DNA bases of the ERE (reviewed in [9, 56]). The so-called D-box in the second motif is important for the contribution of the C-domain to dimerization and consists of a short stretch of amino acids between cysteines 1 and 2 [10]. All three possible dimers of the ER, ERα or ERβ homodimer and ERα/ERβ heterodimer, bind to an identical ERE due to their high homologous DBD (see Fig. 2) [57–59]. Each homodimer and heterodimer can transactivate gene expression. Interestingly, ERβ homodimers appear to be less active than ERα/ERβ heterodimers [17, 57].

5.3
Transcriptional Activation Via AF-1 and AF-2

Data of several laboratories [60–62] revealed that AF-1 functions as a ligand-independent transcription activator, dependent on the cell- and promoter context. These reports also confirm that AF-2 acts in a ligand-dependent manner. Both functions can activate gene expression on their own depending on the cell type and promoter used, as mutational analysis showed. However, full agonistic activity is dependent on both activation functions [63]. A third activation function, termed AF-2a, was discovered by mutational analysis of the human ERα in a yeast system [64]. This AF-2a located within domains D and E (LBD, Fig. 1) was found to be transcriptional, active on its own using a chimeric receptor. Data on ERα/ERβ heterodimers indicate that AF-1 is the dominant activation function and that the AF-2 domains of both ERα and ERβ are necessary for activation [65].

5.4
Co-Activators and -Repressors

After binding its ligand the ER dimer assumes a certain conformation and binds to its ERE. A major question for years dealt with how this ligand-receptor complex induces a differential gene activation or suppression depending on the target cell. It is obvious that the ER transcription complex (ER dimer bound to the ERE) has to interact with the general transcription assembly (GTA) bound to the TATA box within the respective gene promoter (Fig. 4E). The GTA resembles TATA binding proteins (TBP) and various transcription factors (TF) like TFIIA, TFIIB, and TFIID [66] (reviewed in [67–69]). Onate and colleagues [70] were first to clone a co-activator for an SHR. They showed that ER interacts with SRC-1 to induce gene transcription. CBP/p300 is believed to be an essential co-activator interacting with SRC-1 and the GTA [71, 72] (for a review see [73]). Several other co-activators (RIP140, GRIP1 (RIP160 or ERAP-160)) were found and are known to enhance transactivation by ER [73–75] (for reviews see [76, 77]). Recently, an RNA molecule acting as a co-activator, called *steroid receptor RNA activator* or SRA, was discovered [78]. SRA was found to be selective for SHR and to mediate gene activation via the N-terminal AF-1 of the progesterone receptor [78]. To make the ER action even more complex co-repressors were also identified [79, 80]. The co-repressors SMRT and N-CoR were found to inhibit basal promoter activity of the thyroid and retinoic acid receptor when no ligand was bound (for a review see [77]). Montano and colleagues [81] identified a specific repressor of ER gene activation. This co-repressor, termed REA, for *repressor of ER activity*, was also able to inhibit liganded ER and to reverse activation by SRC-1 [81]. The discovery and identification of co-activators and co-repressors provided crucial insights into ER action. The ligand-bound receptor has to be capable of interacting with its co-activators for gene activation and to displace co-repressors to prevent gene silencing. The structural basis of this interaction was provided recently by Shiau and colleagues [52]. They reported the crystal structure of the LBD of human ERα with the agonist diethylstilbestrol (DES) in addition to the binding of the co-activator GRIP1 (Fig. 5). Co-activators like SRC-1 and GRIP1 (ERAP160) bind to a short LXXLL (where L is leucine and X any amino acid) motif in the AF-2 domain, referred to as *nuclear receptor* (NR) box. When DES is bound, helix 12 covers the ligand pocket as described above and allows binding of GRIP1 (Fig. 5). Binding of ER agonists, like DES, induces the conformational changes in the ER necessary for interaction of AF-2 with co-activators leading to full transcriptional activation. This activation takes place depending on the DNA sequence, i.e., the ERE, recognized by the DBD of the ER.

5.5 Estrogen Responsive Elements (EREs)

The ERE is another variable determining ER action. In Fig. 4 (E) the consensus ERE (cERE) is shown, but the ER also recognizes other EREs with base substitutions resulting in an imperfect palindrome (non-consensus ERE, nERE) [82]. Several functional nEREs are known to occur in certain cell-types within different gene promoters [83, 84]. This means that, depending on the presence of an ERE within the regulatory unit of a specific gene present in the target cell, the ER could activate or suppress expression of this gene. However, the presence of an ERE within a gene promoter and the availability of co-factors in a specific cell does not necessarily result in an ER gene regulation. The regulatory DNA sequences are packed into chromatin which restricts the accessibility of the ERE and the gene promoter to the ER and the TFs [82]. For an overview of mechanisms of chromatin remodeling by SHR and histone acetylation, mediated by co-activators like CBP/p300, to allow binding to DNA sequences and gene transcription, see [10, 54, 77, 82, 85–88].

5.6 Cross-Talk with Other Signaling Pathways

The actions of the ER are also modulated by cross-talk with other signaling pathways, like protein kinases (e.g., MAPK) and growth factors (e.g., EGF, IGF-I). This cross-talk modulates not only estrogenic but also anti-estrogenic action, which will be discussed below (for a more detailed discussion see [89–91]).

5.7 Estrogen-Regulated Genes

The expression of a specific gene depends on the accessibility of the ERE, the gene promoter used, and the presence of cell specific co-activators and/or co-repressors. The physiologically observed estrogenic effects are due to this gene activation. Estrogen-regulated genes comprise a wide variety of genes, many of which have unknown activities or importance. Examples of estrogen-responsive genes with known functions are SHR (progesterone receptor), interleukin 4 receptor (part of inflammatory processes), growth factors (EGF, IGF) and growth factor receptors (EGF receptor), cathepsin D (important factor for metastasis of breast tumors), proto-oncogenes (c-myc, c-fos/c-jun), and cell cycle regulatory proteins (cyclin D1) [92–99]. In general, estrogen upregulates cell cycle promoting factors like proto-oncogenes and cyclin D1. The mitogenic effects of estrogenic compounds (e.g., in breast and uterus) are most likely due to these events. The observations that these estrogen-regulated genes are often overexpressed in breast tumors strongly confirm the general notion that activation of the ER is a critical element in tumor progression. Again, it is important to keep in mind that the gene regulation is dependent on the cellular context. This is also illustrated by the induction/suppression of tissue specific genes in bone and heart [100, 101]. In the heart, estrogen is believed to induce various genes (e.g., nitric oxide synthase, myosin) leading to beneficial effects for the cardiovascular system (reviewed in [101]). In bone, estrogen induces osteoblast cell proliferation depending on the differentiation stage, bone matrix protein (e.g., collagen, alkaline phosphatase) synthesis, and growth factors (transforming GFβ (TGFβ), IGFs). These effects lead to improved bone maintenance (reviewed in [100]).

6
Substances Interacting with the ER

Compounds which bind to the ER and induce an ER mediated response, are termed estrogens or ER agonists. Classically, ER antagonists or anti-estrogens are defined as substances preventing or inhibiting the estrogen response. As we have seen in the preceding sections, the ER mediated action is not solely defined by the ER ligand, but rather is the result of the cellular and genetic context. The tissue-dependent activity of ER ligands is highlighted by the drug tamoxifen, used in the treatment of breast cancer, which acts as an ER agonist in the uterus and bone, but as an ER antagonist in the breast. Therefore, the term *selective ER modulators* (SERMs) was introduced [102]. A variety of environmental or industrial chemicals and also naturally occurring substances (phytoestrogens) were found to exert estrogenic or endocrine activity and are referred to as *endocrine disruptors* [1, 2]. These compounds do not comprise a uniformly acting class, but show mixed (estrogenic and/or anti-estrogenic) activities. Since this could potentially lead to beneficial effects (e.g., anti-estrogenic phytoestrogens and the risk of breast cancer [103]) the term *endocrine active compounds* (EACs) was later introduced [104].

Table 1 gives an overview of selected different estrogens, anti-estrogens, SERMs, and EACs.

The most abundant endogenous estrogen is 17β-estradiol (estradiol, Table 1). Synthetic estrogens used as pharmaceutical drugs are the well investigated diethylstilbestrol (DES) and ethinyl estradiol. Tamoxifen is one of the best known and studied SERMs used for breast cancer (antagonistic effects) and more recently also for osteoporosis (agonistic effects) treatment (e.g., [105–108]. Another class are the pure anti-estrogens like ICI 182,780 (ICI) [109–112].

Table 1. Estrogens, antiestrogens, selective estrogen receptor modulators (SERMs), and endocrine active compounds (EACs)

Type of Estrogen	Class	Activity	Compound	Structure
Endogenous estrogen	Steroid	Pure agonist	17*β*-Estradiol (Estradiol)	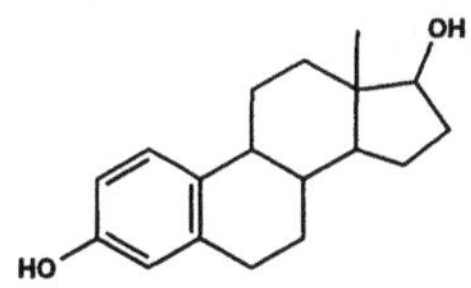
Synthetic "estrogen"	Stilbene	Pure agonist	Diethylstilbestrol (DES)	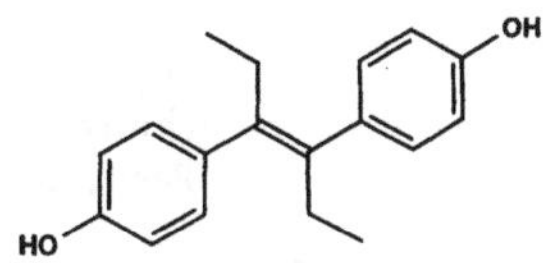
SERM	Non-steroid	Mixed	Tamoxifen	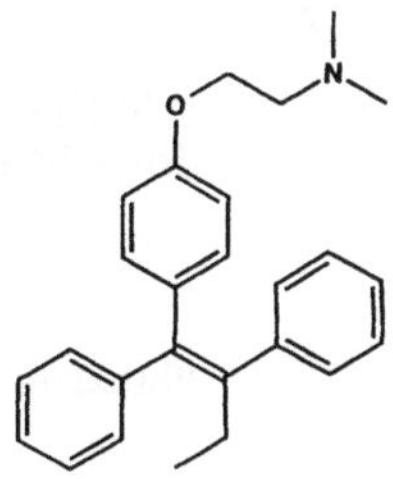
Anti-estrogen	7*α*-substituted Steroid	Pure antagonist	ICI 182,780	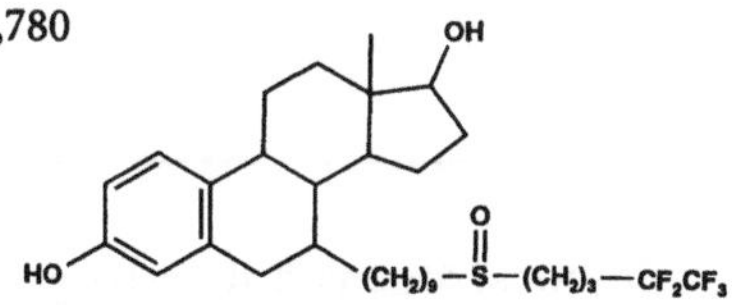
EACs				
Phytoestrogens	Isoflavone	Agonist?/ mixed	Genistein	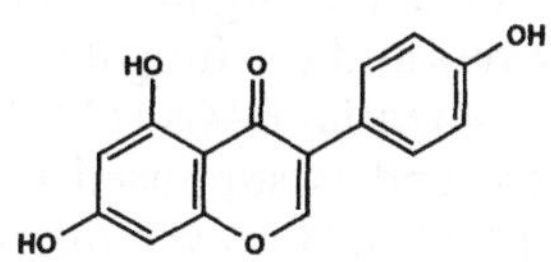
	Lignan	Antagonist	Enterolactone	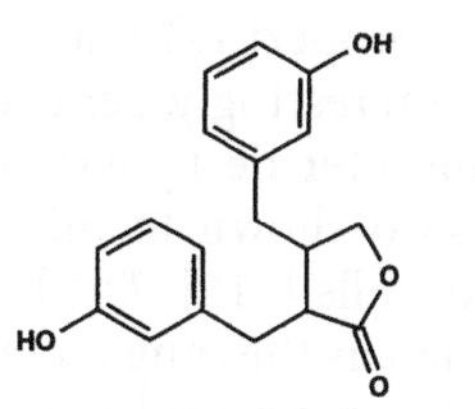
Fungal metabolite	Macrolide	Agonist/ Antagonist	Zearalenone	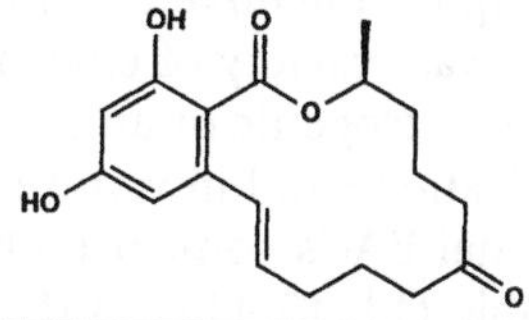

Table 1 (continued)

Type of Estrogen	Class	Activity	Compound	Structure
Environmental chemicals	Phenol	Agonist	Bisphenol A	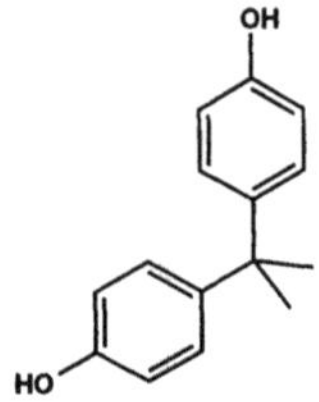
	Organo-chlorines Pesticide	Agonist	*o,p'*-DDT	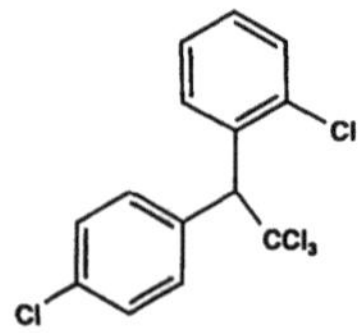
	PCB/Hydroxy-PCB	Agonist/ Antagonist	2',4',6'-tri-chloro-4-biphenylol	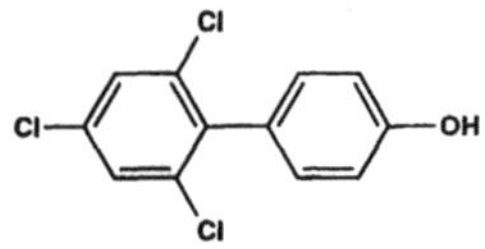
	Phthalate	Agonist	Di-*n*-butyl-phthalate ester	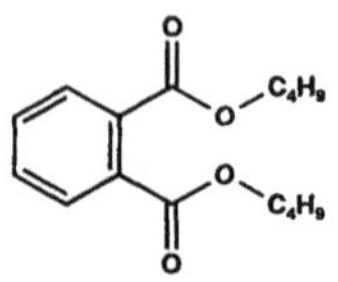

Phytoestrogens/anti-estrogens comprise two main groups: the isoflavones/flavones and lignans (Table 1) [103, 113]. Coumestrol found in lucernes and clovers and the fungal metabolite zearalenone are another class of widely studied potential EACs [113, 114]. Genistein (isoflavone) shows agonistic activity in vitro and is supposed to reduce breast cancer risk by anti-estrogenic action [113, 115]. However, other mechanisms (inhibition of tyrosine kinases, antioxidative properties) may be responsible for the observed cancer preventive effects. Interestingly, genistein acts in mice as an agonist in bone, without inducing any uterine hypertrophy, indicating its potential as a SERM. Isoflavones were also shown to inhibit estrogenicity of pesticides and phenols in mammalian cells [116]. The lignans have been shown to exert antiestrogenic activity, whereas the fungal metabolite zearalenone is mainly an estrogen agonist, although some mixed agonism/antagonism was found via ERβ in vitro [113, 115].

A vast variety of environmental and/or industrial compounds was shown to exert estrogenic or anti-estrogenic effects in vitro (e.g., ER binding, gene transactivation) and in vivo (e.g., uterotrophic effects). These industrial or environmental EACs comprise phenolic compounds, organochlorines, and phthalates (Table 1) [115, 117–119]. Bisphenol A and *o,p'*-DDT showed estrogenic activity in vitro and in vivo [120, 121]. Other phenolic compounds (e.g., nonylphenol, octylphenol) and pesticides (e.g., toxaphene, dieldrin, methoxychlor) also showed weak estrogenic activity in vivo and in vitro [115, 120, 122]. The polychlorinated biphenyls (PCBs) and their hydroxy derivatives (OH-PCBs) are another class of EACs. 2′,4′,6′-Trichloro-4-biphenylol was found to be the most active congener, inducing estrogenic effects in vivo and in vitro [115, 119, 123]. Other OH-PCBs were found to be antiestrogenic [117]. Phthalates are suspected EACs, showing estrogenicity in vitro but not in vivo [124].

The chemicals shown in Table 1 comprise a wide array of compounds with strong agonistic (DES), antagonistic estrogenic activity (ICI 182,780) or mixed activity depending on the target tissue analyzed (tamoxifen). Environmental chemicals show mostly relatively low estrogenic activity (ER binding affinities in vitro $K_d \sim 100-1000$ nmol/l, estradiol ~0.1 nmol/l), whereas the phytoestrogens show stronger, but still low, estrogenicity (isoflavones) and anti-estrogenicity (lignans) ($K_d \sim 10$ nmol/l). Phytoestrogens also show differences in binding and activity for ERα vs ERβ, with a preference for ERβ. Moreover, these phytoestrogens, especially the lignans, occur in much higher abundance than the environmental chemicals depending on the diet.

Not much is known of the mixed activities of the EACs in comparison to the pharmaceutically used SERMs. Often only one estrogenic response (e.g., cell proliferation) and receptor binding was measured in one tissue or cell type. Another important feature that should be taken into account is the activation or deactivation of the parent compound by metabolism in vivo. A weak EAC could be metabolized into a strong EAC with agonistic or antagonist estrogenic properties (e.g., a metabolic activation by hydroxylation is discussed for bisphenol A [121]). This is exemplified for tamoxifen, which itself is a weak SERM. Tamoxifen is metabolized in animals and humans mainly to *N*-desmethyltamoxifen, which has similar estrogenic properties to tamoxifen. Another metabolite is 4-hydroxytamoxifen (OHT, Fig. 6) which shows much higher ER binding affinity and antiestrogenic activity, and can therefore be referred to as the hormonally active metabolite of tamoxifen [125].

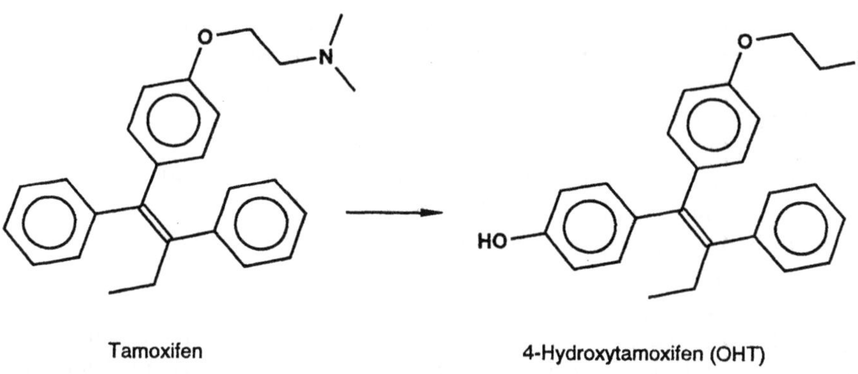

Fig. 6. Metabolism of the SERM tamoxifen results in the active metabolite 4-hydroxytamoxifen (OHT)

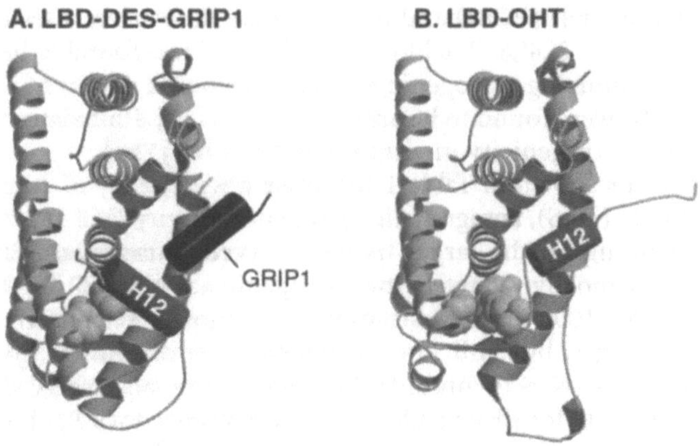

Fig. 7A,B. A Overall structure of the DES-ERα LBD-GRIP1 complex as in Fig. 5. The LBD is shown as ribbon drawing. The co-activator peptide GRIP1 is colored *black* and helix 12 (labeled H12) is colored *gray*, both shown as cylinders. DES is shown in space-filling representation. **B** Overall structure of the OHT-ERα LBD complex. The LBD is depicted as a ribbon drawing. As in A, helix 12 is colored in *gray*. OHT is shown in space-filling representation. The positioning of the OHT sidechain and OHT-mediated structural distortions in and around the ligand-binding pocket stabilizes the conformation of helix 12. Drawn according to [53] using Molscript with the coordinates from the Brookhaven protein databank (accessions 2ERD and 2ERT)

7
Mechanisms of Receptor-Agonists and/or -Antagonists

As we have seen in the preceding subheadings, the binding of a ligand to the ER results in activation (receptor agonism) or in an inhibition of ER mediated action (receptor antagonism) depending on the promoter and cellular context. Since it is well known that compounds like tamoxifen act not as a pure antagonist, but as a selective ER modulator, i.e., as an agonist in uterine, but as an antagonist in breast tissues, one cannot discriminate estrogenic from anti-estrogenic action solely on the ligand structure. However, based on the conformational changes of the ER upon binding of the respective ligand, steroidal estrogens are distinct from SERMs.

Binding of estrogen induces the active conformation of the ER (see Sect. 5). As described above, the binding of the estrogen agonist DES to the LBD of the ER induces helix 12 to reposition over the hydrophobic cavity, activates AF-2, and allows binding of co-activators like GRIP-1 (Fig. 7A and Fig. 5) [52, 53]. OHT, in contrast, occupies more space in the hydrophobic cleft, leading to a distortion of the structure of the LBD (Fig. 7B). The OHT side-chain causes the "misfit" in the binding cavity and forces helix 12 out of its "active" position, preventing binding of GRIP1. The displacement of helix 12 prevents subsequent activation by AF-2. The inhibition of AF-2 is therefore thought to be responsible for the antiestrogenic (antagonistic) activity of OHT [62]. OHT also acts as an agonist, as demonstrated by its uterotrophic action in vivo. Mechanistic studies done in the group of Pierre Chambon revealed that the agonistic activity of

OHT is due to the AF-1 domain in a cell-type and promoter-dependent manner [61]. This suggests that AF-1 is recruiting specific co-activators that are present in the target cell leading to gene activation. Indeed, agonistic response of cells to OHT was enhanced by addition of SRC-1. This activation was independent of AF-2 as shown by AF-2 defective ER mutants [126]. Based on the distinct conformation of the ER induced by each SERM, McDonnell et al. proposed different types of SERMs [63]. OHT belongs to type IV, raloxifene to type III (both classical type I antiestrogens [108]), whereas the pure antiestrogen ICI 182,780 (ICI) is a type II antiestrogen (classical type II antiestrogen [108]). The different, ligand-dependent structures were confirmed just recently using affinity selection of different peptides [127]. The selective peptides were used as probes for different conformations of the ER and showed that estrogen and OHT induce distinct receptor conformations upon binding. The ICI compound also leads to a unique ER structure. Although the pure antagonism of ICI was believed to be due to a complete prevention of the binding of the ER to the ERE, it is known that ICI-bound ER can bind to its ERE, albeit to a reduced extent [63, 128]. However, the transcriptional unit is inactive. Most likely ICI prevents the transport of the newly synthesized receptor to the nucleus and/or induces rapid degradation of the receptor upon ligand binding [108, 112].

The promoter context is another regulating factor. Tzukerman and colleagues [62] showed that tamoxifen acts as an agonist via AF-1 only on certain promoters like the complex complement C3 promoter, whereas the vitellogenin promoter showed activation only when both AF-1 and AF-2 were functional [62, 63]. Different functional non-consensus ERE sequences were found by Dana and colleagues [84]. Interestingly, one of the sequences showed a higher sensitivity to the agonistic activity of tamoxifen than the cERE. A so-called antiestrogen response element (raloxifene response element, RRE) was reported to occur in bone leading to a stronger transcription of the bone-remodeling gene TGFβ3 by raloxifene than by estrogen [129]. Yang and colleagues showed that the isolated RRE may also act in different promoter context, preferentially with raloxifene, although this element alone is not sufficient for full transactivation [129, 130].

To summarize these findings, the proposed mechanistic model includes a distinct shape of the ER upon binding of the ligand. In the case of estrogens, this allows full agonistic activity via AF-1 and AF-2. When SERMs like OHT are bound, AF-2 is inactive resulting in inhibition of ER action. But agonistic activity via AF-1 alone can take place. This AF-1 mediated transactivation depends on the ERE (or antiestrogen response element), the promoter context, if co-repressors can be displaced from the promoter and co- activators recruited in the respective target cell. In contrast, pure antiestrogens, like ICI, prevent any activation via AF-1 or AF-2.

The function of the F-domain of the ER (Fig. 1) is not well defined. Since this carboxy-terminal region has a high interspecies variability, its impact on ER function was supposed to be negligible. However, Montano and colleagues [131] found that the F-domain might modulate the antiestrogenic activity of SERMs like OHT. In human breast cancer (MDA-MB 231) and Chinese hamster ovary (CHO) cells, OHT was able to inhibit estradiol-induced activity by an F-deletion mutant of ER to a much higher extent than by the wild type (F-containing) ER. This F-domain dependent impediment of the antiestrogenic properties of OHT was not seen in other cell types [131].

The activator protein-1 (AP-1) site is one of the best-studied examples for the differential usage of regulatory DNA sequences by ER bound to estrogen or anti-estrogen. AP-1 is the cognate binding site for the transcription factors or proto-oncogenes Fos and Jun and does not represent an ERE. Binding of the heterodimer Fos/Jun can lead to abnormal cell proliferation and transformation and is most likely involved in neoplastic transformation [132, 133]. The AP-1 site is known to be activated by estrogen and the ER [134–136]. Interestingly, tamoxifen and ICI are both able to activate transcription from AP-1 sites via ERα [137, 138]. Tamoxifen showed a higher transactivation capacity in uterine cells than in breast cancer cells. This indicates that the tissue-specific agonistic action known in vivo could be mediated by AP-1 sites [138]. When the same type of experiments were done with ERβ, it was found that the SERMs tamoxifen, raloxifene, and ICI activated AP-1 sites also via ERβ, but estradiol and DES showed no transactivation. Moreover, estradiol and DES inhibited the SERM induced ERβ-dependent transcription from the AP-1 site [137]. Notably, the ICI compound, which is considered to be a pure antiestrogen without estrogenic activity in any tissue yet investigated, activated transcription via ER. This transactivational capability of ICI was supported by studies on DNA binding of ER [58]. Pace and colleagues [58] showed that ICI stabilized the binding of ERβ to DNA, but not of ERα. The approach of Paige and colleagues [127] confirming the different receptor conformation of estrogen or OHT complexed ER, as described above, also revealed differences between ERα and ERβ. The two ER subtypes showed structural differences when bound to the same ligand [127]. These studies highlighted the potential of differential action of estrogens and SERMs via ERα and ERβ. Indeed, several SERMs and ICI showed differential binding affinities to ERα and ERβ and also different transactivational capacities [139]. The potential of ERβ to mediate different or even antagonistic effects when compared to ERα may also play a role in regard of EACs. Kuiper and colleagues [29] found for some phytoestrogens, like genistein, higher binding affinities to ERβ than to ERα. Accordingly, they also found stronger inhibition of estradiol action on ERβ than on ERα [29].

Another important factor for discrimination between estrogens and SERMs or possibly EACs is cross-talk with other signaling pathways. Work from the group of Benita Katzenellenbogen showed that enhancement of protein kinase activity synergizes with ER mediated gene transcription (reviewed in [89, 140]). This synergism can be abolished by pure antiestrogens, like ICI [140]. Kinase pathways can also elevate agonistic action of tamoxifen and reduce its antagonistic activity, whereas no effect was seen for ICI [141]. These results indicate that the described cross-talk of ER action could modulate or possibly reverse the activity of SERMs. Estradiol also activates another mitogenic pathway, the tyrosine kinase/p21^{ras}/mitogen activated protein kinase (MAPK) signal transduction cascade, which plays a major role in cell proliferation [142]. These proliferation-enhancing pathways can be inhibited by ICI. Tamoxifen and ICI also inhibit the mitogenic effects of estrogens via growth factors TGFα and IGF-I and cell cycle regulatory proteins (cyclin D1) (reviewed in [108]). In breast cancer cells, tamoxifen and ICI inhibited IGF-I action [143, 144], whereas only tamoxifen induced IGF-I in the uterus, which may then be responsible for its uterotrophic response [145]. This emphasizes that differential cross-talk with other signaling pathways could lead to a tissue-specific response of SERMs. The above-mentioned induction of AP-1 activity by estrogens was seen in ER positive human breast cancer cells, but not in other cell types transiently transfected with ER. Tamoxifen and ICI inhibited activation of growth factors via ER and the AP-1 site in breast cancer cells [146]. These reports suggest that SERMs can antagonize the mitogenic effects of estrogen in concert with other signal transduction cascades in a cell-type specific manner.

Interaction of target cells with SERMs could also lead to a variety of effects not directly mediated by the ER. These include degradation or loss of the receptor and could induce resistance to antagonistic activity of SERMs unfortunately seen during the treatment of breast cancer with antiestrogens (for a review see [108, 147]). Studies in vitro showed that cell lines growing in estrogen-free medium for an extended period of time are losing functional ER and grow maximally without estrogen [148–151]. The SERM tamoxifen stabilizes ER protein in cells, whereas ICI reduces ER protein levels, potentially leading to a loss of ER and estrogen resistance [152].

The preceding sections showed that the actions of estrogens, anti-estrogens, SERMs, or EACs are a complex array of the ER signal transduction and its interaction with other cell signaling pathways. Katzenellenbogen and colleagues [153] have proposed the *tripartite model* of receptor action. In this model, the ligand-receptor complex defines the efficacy of the interaction with the so-

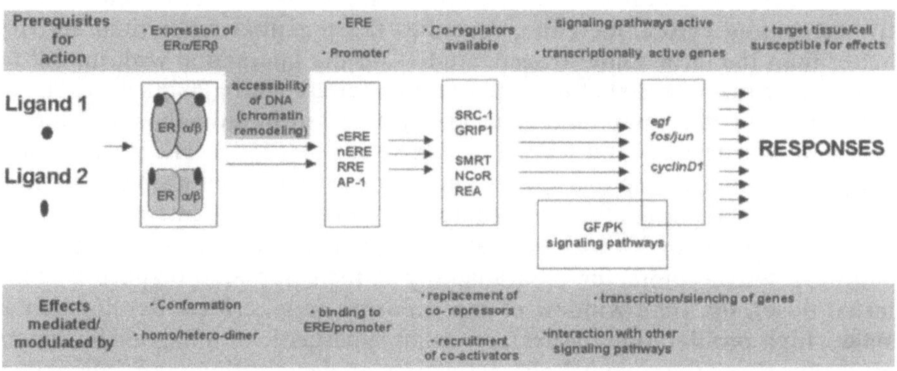

Fig. 8. Schematic illustration of the complexity of ER mediated responses induced by different ligands. Examples of the different variables are given in the boxes. On the *top*, the general prerequisites, depending on the target cell or tissue, for ligand-induced ER action are shown. How this action is mediated and modulated by is shown at the *bottom*. *Arrows* between the boxes indicate the various possibilities of actions leading to different responses mediated by the ligand bound to the ER. cERE, consensus ERE; nERE, non-consensus ERE; RRE, raloxifen RE; AP-1, activator protein-1 binding site; SRC-1/GRIP1, co-activators; SMRT/NcoR/REA, co-repressors; GF, growth factor; PK, protein kinase; *egf*, epidermal growth factor gene. See text for details

called *effector* system. The receptor-*effector* complex then defines the responses or effects. The coupling of the ligand-receptor complex with the *effector* system comprises all modulating factors shown in Fig. 8. This scheme emphasizes the variety of factors modulating ER action induced by different ligands. The respective ligand forces the receptor into a specific conformation. Depending on the ER subtype present in the target cell, the accessibility of the regulatory DNA sequences containing cERE, nERE, or antiestrogen response elements (e.g., RRE) and the promoter being present, the active ER (as a homo- and/or heterodimer) has to replace co-repressors and recruit co-activators to transcribe a specific gene. All these modulating factors are dependent on the cellular context and can be enhanced or inhibited by cross-talk with other signaling pathways. The proteins expressed as a result of this complex concert then exert their effects on the target cell and consequently define the tissue response (Fig. 8).

In conclusion, the potency of the effect of the ER action induced by a specific ligand, estrogen, antiestrogen, SERM, or EAC is not solely defined by the ligand structure but rather by the interaction of the receptor-ligand complex with various modulating factors. The interplay between ligand bound ER and these factors determines the response depending on the target cell.

8
Endocrine Active Compounds: Estrogens or Antiestrogens, or Both?

A vast variety of xenobiotics has been shown to act via the ER. The complex action of well studied compounds like tamoxifen via the ER, acting as estrogen and anti-estrogen, exemplifies that the effects of one xenobiotic cannot be deduced from its structure or one functional assay alone. The term *endocrine active compound* (EAC), therefore, describes the potential differential activity better than the term xenoestrogen. Studies on the interaction with the ER in vitro define the potential of a compound to interfere with ER signaling. The in vivo effects of this compound can then be assessed in animal assays (for an overview of screening and testing of EACs see [154, 155]). The potency of a particular compound to disrupt the endocrine system is not a single function of its receptor binding affinity, but rather is defined on the interaction of the ligand-receptor complex with the transcription machinery and other signaling pathways. Furthermore, the bioavailability of the compound (exposure vs internal dose), the time window of the exposure (e.g., critical developmental stages, high proliferating tissues), potential synergistic or antagonistic effects of chemical mixtures, metabolic activation or deactivation, and interaction with other endocrine systems or receptors, such as Ah-receptor, retinoic acid receptor, or androgen receptor, will certainly modulate the compound's physiological effects (for a discussion of doses and physiological effects of EACs see [104, 156]).

It is therefore necessary to study thoroughly the cellular and molecular interactions with the ER, regarding different target tissues, level of ER expression, cell signaling pathways, and target genes in vitro and in vivo to assess the potential of a chemical to disrupt ER mediated signaling.

Acknowledgments. This work was supported in part by a Grant of the Deutsche Forschungsgemeinschaft (Mu 1490/1) to S.O.M. We gratefully acknowledge Dr. Lars Pedersen for constructing the structures in Figs. 5 and 7, and Dr. Edina Burns and Bonnie Deroo for critical and careful editing of the manuscript.

9 References

1. Colburn T, Clement C (1992) Chemically induced alterations in sexual and functional development: the wildlife/human connection. Princeton Scientific Publishing, Princeton
2. McLachlan JA, Korach KS (1995) Environ Health Perspect 103(suppl 7):3
3. Newbold RR, McLachlan JA (1996) Transplacental hormonal carcinogenesis: diethylstilbestrol as an example. In: Huff J, Boyd J, Barrett JC (eds) Cellular and molecular mechanisms of hormonal carcinogenesis: environmental influences. Wiley-Liss, New York, p 131
4. Griffin JE, Ojeda SR (1996) Textbook of endocrine physiology. Oxford University Press, New York
5. Lindzey J, Korach KS (1997) Steroid hormones. In: Conn PM, Melmed S (eds) Endocrinology: basic and clinical principles. Humana Press, Totowa, p 47
6. Kliewer SA, Lehmann JM, Willson TM (1999) Science 284:757
7. Evans RM (1988) Science 240:889
8. Mangelsdorf DJ, Thummel C, Beato M, Herrlich P, Schuetz G, Umesono K, Blumberg B, Kastner P, Mark M, Chambon P, Evans RM (1995) Cell 83:835
9. Tsai M-J, O'Malley B (1994) Annu Rev Biochem 63:451
10. Beato M, Herrlich P, Schuetz G (1995) Cell 83:851
11. Greene GL, Gilna P, Waterfield M, Baker A, Hort Y, Shine J (1986) Science 231:1150
12. Green S, Walter P, Kumar V, Krust A, Bornert JM, Argos P, Chambon P (1986) Nature 320:134
13. Kuiper GGJM, Enmark E, Pelto-Huiko M, Nilsson S, Gustafsson J-A (1996) Proc Natl Acad Sci USA 93:5925
14. Mosselman S, Polman J, Dijkema R (1996) FEBS Lett 392:49
15. White R, Lees JA, Needham M, Ham J, Parker M (1987) Mol Endocrinol 1:735
16. Enmark E, PeltoHuikko M, Grandien K, Lagercrantz S, Lagercrantz J, Fried G, Nordenskjold M, Gustafsson JA (1997) J Clin Endocrinol Metab 82:4258
17. Ogawa S, Inoue S, Watanabe T, Hiroi H, Orimo A, Hosoi T, Ouchi Y, Muramatsu M (1998) Biochem Biophys Res Commun 243:122
18. Leygue E, Dotzlaw H, Lu B, Glor C, Watson PH, Murphy LC (1998) J Clin Endocrinol Metab 83:3754
19. Moore JT, McKee DD, Slentz Kesler K, Moore LB, Jones SA, Horne EL, Su JL, Kliewer SA, Lehmann JM, Willson TM (1998) Biochem Biophys Res Commun 247:75
20. Fuqua SA, Chamness GC, McGuire WL (1993) J Cell Biochem 51:135
21. Fuqua SA, Wolf DM (1995) Breast Cancer Res Treat 35:233
22. Hopp TA, Fuqua SAW (1998) J Mammary Gland Biol Neopl 3:73
23. Leygue ER, Watson PH, Murphy LC (1996) J Natl Cancer Inst 88:284
24. Murphy LC, Dotzlaw H, Leygue E, Douglas D, Coutts A, Watson PH (1997) J Steroid Biochem Mol Biol 62:363
25. Petersen DN, Tkalcevic GT, KozaTaylor PH, Turi TG, Brown TA (1998) Endocrinol 139:1082
26. Encarnacion CA, Fuqua SA (1994) Cancer Treat Res 71:97
27. Leygue E, Dotzlaw H, Watson PH, Murphy LC (1999) Cancer Res 59:1175
28. Catherino WH, Wolf DM, Jordan VC (1995) Mol Endocrinol 9:1053
29. Kuiper GGJM, Carlsson B, Grandien K, Enmark E, Haggblad J, Nilsson S, Gustafsson J-A (1997) Endocrinol 138:863
30. Couse JF, Lindzey JK, Grandien K, Gustafsson J-A, Korach KS (1997) Endocrinol 138:4613

31. Shughure PJ, Lane MV, Scrimo PJ, Merchenthaler I (1998) Steroids 63:498
32. Saunders PTK, Fisher JS, Sharpe RM, Millar MR (1998) J Endocrinol 156:R13
33. Pedeutour F, Quade BJ, Weremowicz S, DalCin P, Ali S, Morton CC (1998) Gene Chromosome Cancer 23:361
34. Sar M, Welsch F (1999) Endocrinol 140:963
35. Mitchner NA, Garlick C, Ben-Jonathan N (1998) Endocrinol 139:3976
36. Ankrom MA, Patterson JA, Davis PY, Vetter UK, Blackman MR, Sponseller PD, Tayback M, Robey PG, Shapiro JR, Fedarko NS (1998) Biochem J 333:787
37. Nilsson LO, Boman A, Savendahl L, Grigelioniene G, Ohlsson C, Ritzen EM, Wroblewski J (1999) J Clin Endocrinol Metab 84:370
38. Onoe Y, Miyaura C, Ohta H, Nozawa S, Suda T (1997) Endocrinol 138:4509
39. Register TC, Adams MR (1998) J Steroid Biochem Mol Biol 64:187
40. Ciocca DR, Vargas Roig LM (1995) Endocrine Rev 16:35
41. Couse JF, Korach KS (1999) Endocrine Rev 20:358
42. Lubahn DB, Moyer JS, Golding TS, Couse JF, Korach KS, Smithies O (1993) Proc Natl Acad Sci USA 90:11,162
43. Korach KS, Couse JF, Curtis SW, Wasburn TF, Lindzey JK, Kimbro KS, Eddy EM, Migliaccio S, Snedecker SM, Lubahn DB, Schomberg DW, Smith EP (1996) Recent Prog Horm Res 51:159
44. Krege JH, Hodgin JB, Couse JF, Enmark E, Warner M, Mahler JF, Sar M, Korach KS, Gustafsson JA, Smithies O (1998) Proc Natl Acad Sci USA 95:15,677
45. Knobil E, Neill JD (1994) The physiology of reproduction, vols 1 and 2. Raven Press, New York
46. Buchner J (1999) Trends Biochem Sci 24:136
47. Smith DF, Toft DO (1993) Mol Endocrinol 7:4
48. Washburn T, Hocutt A, Brautigan DL, Korach KS (1991) Mol Endocrinol 5:235
49. Orti E, Bodwell JE, Munck A (1992) Endocrine Rev 13:105
50. Bunone G, Briand PA, Miksicek RJ, Picard D (1996) EMBO J 15:2174
51. Kato S, Endoh H, Masuhiro Y, Kitamoto T, Uchiyama S, Sasaki H, Masushige S, Gotoh Y, Nishida E, Kawashima H, Metzger D, Chambon P (1995) Science 270:1491
52. Shiau AK, Barstad D, Loria PM, Cheng L, Kushner PJ, Agard DA, Greene GL (1998) Cell 95:927
53. Brzozowski AM, Pike AC, Dauter Z, Hubbard RE, Bonn T, Engstrom O, Ohman L, Greene GL, Gustafsson JA, Carlquist M (1997) Nature 389:753
54. Moras D, Gronemeyer H (1998) Curr Opin Cell Biol 10:384
55. Katzenellenbogen BS, Bhardwaj B, Fang H, Ince BA, Pakdel F, Reese JC, Schodin D, Wrenn CK (1993) J Steroid Biochem Mol Biol 47:39
56. Freedman LP, Luisi BF (1993) J Cell Biochem 51:140
57. Cowley SM, Hoare S, Mosselman S, Parker MG (1997) J Biol Chem 272:19,858
58. Pace P, Taylor J, Suntharalingam S, Coombes RC, Ali S (1997) J Biol Chem 272:25,832
59. Pettersson K, Grandien K, Kuiper GG, Gustafsson JA (1997) Mol Endocrinol 11:1486
60. Ignar-Trowbridge DM, Teng CT, Ross KA, Parker MG, Korach KS, McLachlan JA (1993) Mol Endocrinol 7:992
61. Berry M, Metzger D, Chambon P (1990) EMBO J 9:2811
62. Tzukerman MT, Esty A, Santiso-Mere D, Danielian P, Parker MG, Stein RB, Pike JW, McDonnell DP (1994) Mol Endocrinol 8:21
63. McDonnell DP, Dana SL, Hoener PA, Lieberman BA, Imhof MO, Stein RB (1995) Ann NY Acad Sci 761:121
64. Norris JD, Fan D, Kerner SA, McDonnell DP (1997) Mol Endocrinol 11:747
65. Tremblay GB, Tremblay A, Labrie F, Giguere V (1999) Mol Cell Biol 19:1919
66. Ing NH, Beekman JM, Tsai SY, Tsai MJ, O'Malley BW (1992) J Biol Chem 267:17,617
67. Beato M, Sanchez Pacheco A (1996) Endocrine Rev 17:587
68. Roeder RG (1996) Trends Biochem Sci 21:327
69. Roeder RG (1998) Cold Spring Harb Symp Quant Biol 63:201
70. Onate SA, Tsai SY, Tsai MJ, O'Malley BW (1995) Science 270:1354

71. Chakravarti D, LaMorte VJ, Nelson MC, Nakajima T, Schulman IG, Juguilon H, Montminy M, Evans RM (1996) Nature 383:99
72. Hanstein B, Eckner R, DiRenzo J, Halachmi S, Liu H, Searcy B, Kurokawa R, Brown M (1996) Proc Natl Acad Sci USA 93:11,540
73. Glass CK, Rose DW, Rosenfeld MG (1997) Curr Opin Cell Biol 9:222
74. Halachmi S, Marden E, Martin G, MacKay H, Abbondanza C, Brown M (1994) Science 264:1455
75. Cavailles V, Dauvois S, L'Horset F, Lopez G, Hoare S, Kushner PJ, Parker MG (1995) EMBO J 14:3741
76. Shibata H, Spencer TE, Onate SA, Jenster G, Tsai SY, Tsai MJ, O'Malley BW (1997) Recent Prog Horm Res 52:141
77. McKenna NJ, Lanz RB, O'Malley BW (1999) Endocrine Rev 20:321
78. Lanz RB, McKenna NJ, Onate SA, Albrecht U, Wong J, Tsai SY, Tsai M-J, O'Malley BW (1999) Cell 97:17
79. Chen JD, Evans RM (1995) Nature 377:454
80. Horlein AJ, Naar AM, Heinzel T, Torchia J, Gloss B, Kurokawa R, Ryan A, Kamei Y, Soderstrom M, Glass CK, Rosenfeld MG (1995) Nature 377:397
81. Montano MM, Ekena K, Delage-Mourroux R, Chang WR, Martini P, Katzenellenbogen BS (1999) Proc Natl Acad Sci USA 96:6947
82. Truss M, Beato M (1993) Endocrine Rev 14:459
83. Klein Hitpass L, Schorpp M, Wagner U, Ryffel GU (1986) Cell 46:1053
84. Dana SL, Hoener PA, Wheeler DA, Lawrence CB, McDonnell DP (1994) Mol Endocrinol 8:1193
85. Wolffe AP (1997) Cell Res 7:127
86. Archer TK, Zaniewski E, Moyer ML, Nordeen SK (1994) Mol Endocrinol 8:1154
87. Hager GL, Archer TK, Fragoso G, Bresnick EH, Tsukagoshi Y, John S, Smith CL (1993) Cold Spring Harb Symp Quant Biol 58:63
88. Mymryk JS, Fryer CJ, Jung LA, Archer TK (1997) Methods 12:105
89. Katzenellenbogen BS (1996) Biol Reprod 54:287
90. Ignar Trowbridge DM, Pimentel M, Teng CT, Korach KS, McLachlan JA (1995) Environ Health Perspect 7:35
91. Smith CL (1998) Biol Reprod 58:627
92. van der Burg B, de Groot RP, Isbrucker L, Kruijer W, de Laat SW (1992) J Steroid Biochem Mol Biol 43:111
93. Rivera Gonzalez R, Petersen DN, Tkalcevic G, Thompson DD, Brown TA (1998) J Steroid Biochem Mol Biol 64:13
94. De Bortoli M, Dati C, Antoniotti S, Maggiora P, Sapei ML (1992) J Steroid Biochem Mol Biol 43:21
95. Rochefort H (1995) Ciba Found Symp 191:254
96. Rochefort H, Capony F, Garcia M (1990) Cancer Metastasis Rev 9:321
97. Prall OW, Rogan EM, Sutherland RL (1998) J Steroid Biochem Mol Biol 65:169
98. Weisz A, Bresciani F (1993) Crit Rev Oncog 4:361
99. May FE, Westley BR (1995) Biomed Pharmacother 49:400
100. Oursler MJ (1998) Crit Rev Eukaryot Gene Expr 8:125
101. Pelzer T, Shamim A, Neyses L (1996) Mol Cell Biochem 160/161:307
102. Mitlak BH, Cohen FJ (1997) Horm Res 48:155
103. Adlercreutz H (1995) Environ Health Perspect 103:103
104. Barton HA, Andersen ME (1998) Crit Rev Toxicol 28:363
105. Jordan VC (1998) J Natl Cancer Inst 90:967
106. Jordan VC, Morrow M (1999) Endocrine Rev 20:253
107. Cosman F, Lindsay R (1999) Endocrine Rev 20:418
108. MacGregor JI, Jordan VC (1998) Pharmacol Rev 50:151
109. Gradishar WJ, Jordan VC (1999) Endocrinology of Breast Cancer 11:283
110. Wakeling AE, Bowler J (1992) J Steroid Biochem Mol Biol 43:173
111. Wakeling AE, Dukes M, Bowler J (1991) Cancer Res 51:3867

112. Gibson MK, Nemmers LA, Beckman WC Jr, Davis VL, Curtis SW, Korach KS (1991) Endocrinol 129:2000
113. Bingham SA, Atkinson C, Liggins J, Bluck L, Coward A (1998) Br J Nutr 79:393
114. Jordan VC, Mittal S, Gosden B, Koch R, Lieberman ME (1985) Environ Health Perspect 61:97
115. Kuiper G, Lemmen JG, Carlsson B, Corton JC, Safe SH, van der Saag PT, van der Burg P, Gustafsson JA (1998) Endocrinol 139:4252
116. Verma SP, Goldin BR, Lin PS (1998) Environ Health Perspect 106:807
117. Safe S, Connor K, Ramamoorthy K, Gaido K, Maness S (1997) Regul Toxicol Pharmacol 26:52
118. Safe SH (1995) Environ Health Perspect 103:346
119. Korach KS, Sarver P, Chae K, McLachlan JA, McKinney JD (1988) Mol Pharmacol 33:120
120. Shelby MD, Newbold RR, Tully D, Chae K, Davis VL (1996) Environ Health Perspect 104:1296
121. Ben-Jonathan N, Steinmetz R (1998) Trends Endocrinol Metab 9:124
122. Ramamoorthy K, Wang F, Chen I-C, Norris JD, McDonnell DP, Leonard L, Gaido K, Bocchinofuso WP, Korach KS, Safe S (1997) Endocrinol 138:1520
123. Ramamoorthy K, Vyhlidal C, Wang F, Chen IC, Safe S, McDonnell DP, Leonard LS, Gaido KW (1997) Tox Appl Pharmacol 147:93
124. Zacharewski TR, Meek MD, Clemons JH, Wu ZF, Fielden MR, Matthews JB (1998) Toxicol Sci 46:282
125. Jordan VC, Robinson SP (1987) Fed Proc 46:1870
126. Smith CL, Nawaz Z, O'Malley BW (1997) Mol Endocrinol 11:657
127. Paige LA, Christensen DJ, Gron H, Norris JD, Gottlin EB, Padilla KM, Chang CY, Ballas LM, Hamilton PT, McDonnell DP, Fowlkes DM (1999) Proc Natl Acad Sci USA 96:3999
128. Curtis SW, Korach KS (1991) Mol Endocrinol 5:959
129. Yang NN, Venugopalan M, Hardikar S, Glasebrook A (1996) Science 273:1222
130. Yang NN, Venugopalan M, Hardikar S, Glasebrook A (1997) Science 275:1249
131. Montano MM, Mueller V, Trobaugh A, Katzenellenbogen BS (1995) Mol Endocrinol 9:814
132. Cohen DR, Curran T (1989) Crit Rev Oncog 1:65
133. Curran T, Franza BR Jr (1988) Cell 55:395
134. Umayahara Y, Kawamori R, Watada H, Imano E, Iwama N, Morishima T, Yamasaki Y, Kajimoto Y, Kamada T (1994) J Biol Chem 269:16,433
135. Philips A, Chalbos D, Rochefort H (1993) J Biol Chem 268:14,103
136. Gaub MP, Bellard M, Scheuer I, Chambon P, Sassone Corsi P (1990) Cell 63:1267
137. Paech K, Webb P, Kuiper GG, Nilsson S, Gustafsson J, Kushner PJ, Scanlan TS (1997) Science 277:1508
138. Webb P, Lopez GN, Uht RM, Kushner PJ (1995) Mol Endocrinol 9:443
139. Barkhem T, Carlsson B, Nilsson Y, Enmark E, Gustafsson J, Nilsson S (1998) Mol Pharmacol 54:105
140. Cho H, Katzenellenbogen BS (1993) Mol Endocrinol 7:441
141. Fujimoto N, Katzenellenbogen BS (1994) Mol Endocrinol 8:296
142. Migliaccio A, Di Domenico M, Castoria G, de Falco A, Bontempo P, Nola E, Auricchio F (1996) EMBO J 15:1292
143. Winston R, Kao PC, Kiang DT (1994) Breast Cancer Res Treat 31:107
144. Freiss G, Rochefort H, Vignon F (1990) Biochem Biophys Res Commun 173:919
145. Huynh HT, Pollak M (1993) Cancer Res 53:5585
146. Chalbos D, Philips A, Rochefort H (1994) Semin Cancer Biol 5:361
147. Katzenellenbogen BS, Montano MM, Ekena K, Herman ME, McInerney EM (1997) Breast Cancer Res Treat 44:23
148. Cho HS, Ng PA, Katzenellenbogen BS (1991) Mol Endocrinol 5:1323
149. Clarke R, Brunner N, Katzenellenbogen BS, Thompson EW, Norman MJ, Koppi C, Paik S, Lippman ME, Dickson RB (1989) Proc Natl Acad Sci USA 86:3649
150. Murphy CS, Meisner LF, Wu SQ, Jordan VC (1989) Eur J Cancer Clin Oncol 25:1777

151. Pink JJ, Bilimoria MM, Assikis J, Jordan VC (1996) Br J Cancer 74:1227
152. Pink JJ, Jordan VC (1996) Cancer Res 56:2321
153. Katzenellenbogen JA, O'Malley BW, Katzenellenbogen BS (1996) Mol Endocrinol 10:119
154. Safe S, Connor K, Gaido K (1998) Toxicol Lett 103:665
155. Gray LE Jr, Kelce WR, Wiese T, Tyl R, Gaido K, Cook J, Klinefelter G, Desaulniers D, Wilson E, Zacharewski T, Waller C, Foster P, Laskey J, Reel J, Giesy J, Laws S, McLachlan J, Breslin W, Cooper R, Di Giulio R, Johnson R, Purdy R, Mihaich E, Safe S, Colborn T (1997) Reprod Toxicol 11:719
156. Golden RJ, Noller KL, Titusernstoff L, Kaufman RH, Mittendorf R, Stillman R, Reese EA (1998) Crit Rev Toxicol 28:109

CHAPTER 2

In Vitro Methods for Characterizing Chemical Interactions with Steroid Hormone Receptors

Kevin W. Gaido[1], Donald P. McDonnell[2], Stephen Safe[3]

[1] CIIT Centers for Health Research, Research Triangle Park, NC 27709, USA
e-mail: gaido@ciit.org

[2] Department of Pharmacology and Cancer Biology, Duke University Medical Center, Durham, NC 27710, USA

[3] Department of Veterinary Physiology and Pharmacology, Texas A&M University, College Station, TX 77843-4466, USA

In response to concern over possible human exposure to endocrine active chemicals the USEPA has proposed an Endocrine Disruptor screening program to evaluate endocrine-disrupting properties of chemical substances and common mixtures. The program will evaluate tens of thousands of compounds and as a result assays will need to be developed that will allow for the screening of a large number of chemicals. This chapter describes several in vitro assays for characterizing chemical interaction with steroid hormone receptors. These assays are relatively rapid and inexpensive, and they allow for the testing of multiple chemicals over a wide dose range. Some of these assays are likely to be considered by the USEPA as part of a screening strategy for endocrine active chemicals. In vitro assays can also be used to obtain mechanistic information that can aid in understanding and predicting the in vivo effects of endocrine active chemicals.

Keywords. Steroid hormone receptor, Endocrine disruptors, In vitro assays, Bisphenol A, Methoxychlor

1 **Introduction** 28

2 **Competitive Binding Assays** 28

3 **Yeast-Based Steroid Receptor Assays** 30

4 **Mammalian-Based Reporter Gene Assays** 31

5 **Cell Proliferation Assays** 33

6 **Peptide Probes** 34

7 **Summary** 35

8 **References** 36

The Handbook of Environmental Chemistry Vol. 3, Part L
Endocrine Disruptors, Part I
(ed. by M. Metzler)

List of Abbreviations

BPA	bisphenol A
E2	17β-estradiol
EDSP	Endocrine Disruptor Screening Program
ER	estrogen receptor
HPTE	2,2-bis(*p*-hydroxyphenyl)-1,1,1-trichloroethane
HRP	horseradish peroxidase
RBA	relative binding affinity
USEPA	United States Environmental Protection Agency

1 Introduction

In response to concern over possible human exposure to endocrine active chemicals, changes were made in 1996 to the Food Quality Protection Act and the Safe Drinking Water Act which mandated the United States Environmental Protection Agency (USEPA) to develop a screening and testing strategy for endocrine active chemicals by August 1998 and implement the plan by August, 1999. USEPA has proposed an Endocrine Disruptor screening program (EDSP) to evaluate endocrine-disrupting properties of chemical substances and common mixtures. The program will evaluate more than 87,000 industrial chemicals including commercial chemicals, active pesticides ingredients, ingredients in cosmetics, nutritional supplements, and food additives. The enormous scope of this program necessitates a tiered approach that will include priority setting, Tier 1 screening, and Tier 2 testing. The priority setting and Tier 1 stages will include in vitro tests designed to identify chemicals capable of interacting with steroid hormone receptors.

In vitro assays for characterizing chemical interaction with steroid hormone receptors include receptor binding, cell proliferation, and reporter gene assays. Many of these in vitro assays can be adapted for use as screening tools and some are likely to be considered by the USEPA as part of a screening strategy for endocrine active chemicals. Moreover, in addition to their utility to screen chemicals for interactions with steroid hormone receptors, in vitro assays can be used to obtain mechanistic information that can aid in understanding and predicting the in vivo effects of endocrine active chemicals.

2 Competitive Binding Assays

A competitive binding assay measures the binding of a single concentration of radiolabeled ligand in the presence of various concentrations of competing unlabeled ligand [1–3]. This is a rapid and inexpensive method for determining whether a chemical interacts with the hormone-binding pocket (site) of a specific steroid hormone receptor [4–7]. Hormone receptor preparations can be obtained from various tissue or whole cell homogenates, or as purified recombinant protein. The receptor is incubated with radiolabeled ligand in the presence or absence of increasing concentrations of unlabeled test chemical. If the chemical competes with the radiolabeled ligand for receptor binding it will displace a fraction of the radiolabeled ligand from the receptor in a concentration dependent manner. Receptor bound radiolabeled ligand is separated from unbound ligand by filtration, hydroxyapatite extraction, or other methods and quantified by scintillation counting. Data can be analyzed by nonlinear regression to obtain an IC_{50}, or the concentration of unlabeled ligand that displaces 50% of the bound radioligand from the receptor. The relative binding affinity (RBA) is determined by dividing the IC_{50} value for a test chemical by the IC_{50} value for a standard receptor ligand or the radioligand itself. RBA values of test chemicals can then be compared as a measure of their relative binding affinity for the receptor. It is important to remember that the IC_{50} value is a property of the experiment. By changing experimental conditions (such as changing the radioligand or its concentration, or changing the source of the receptor preparation) you can change the IC_{50}.

A competitive binding assay comparing the relative estrogen receptor binding affinity of estradiol (E2) and bisphenol A (BPA), a monomer used in the production of polycarbonate and epoxy resins, and 2,2-bis(*p*-hydroxyphenyl)-1,1,1-trichloroethane (HPTE), a metabolite of the pesticide methoxychlor, is presented in Fig. 1. The results show that BPA and HPTE have an approximately 3000-fold and 95-fold lower binding affinity for the estrogen receptor (ER), respectively, than does E2.

A method that uses fluorescence polarization to monitor displacement of a fluorescent ligand has recently been used to characterize chemical interaction with the ER [8]. Fluorescent polarization is a measure of the speed of rotation of a fluorescent molecule. When a fluorescently labeled ligand binds to a receptor its size and rotation are altered. The polarization value is converted into the concentration of bound ligand and the resulting binding data are analyzed as described for traditional competitive binding assays. This method offers several advantages over traditional binding assays in that it does not require radioac-

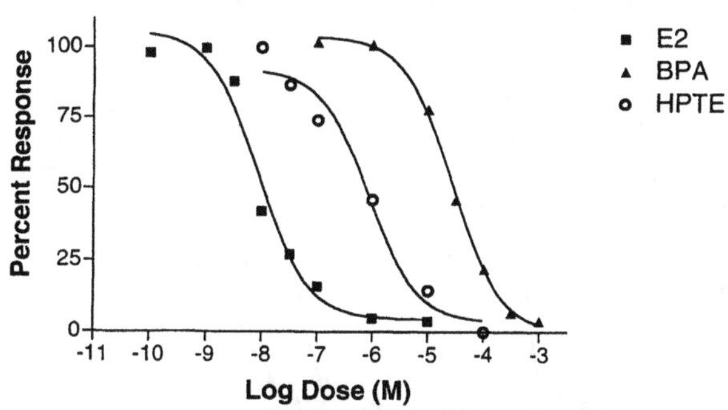

Fig. 1. Competitive binding assay comparing the relative binding of estradiol and bisphenol A to the purified recombinant human estrogen receptor. Estradiol, HPTE, and bisphenol A were combined with 5 nmol/l [3H]-estradiol and recombinant human estrogen receptor. After overnight incubation the receptor complex was precipitated with hydroxyapatite and the radioactivity of the pellets was determined using a scintillation counter

tivity, no separation step is required since the method gives a direct measure of bound versus free fluorescent ligand, and binding reaction kinetics can be monitored.

Competitive binding assays are quick, sensitive, and specific, and they allow for the testing of a chemical over a wide dose range. However, competitive binding assays do not distinguish between receptor agonists and antagonists. Thus, from the results presented in Fig. 1 it cannot be determined whether BPA or HPTE are ER agonists or antagonists. Competitive binding assays may also yield false negative results if metabolic activation is required before binding to the steroid receptor. In addition, results may be artifactual if the receptor is altered by detergent or denaturation effects of a test chemical.

3 Yeast-Based Steroid Receptor Assays

Yeast do not contain endogenous steroid hormone receptors; however, mammalian steroid hormone receptors introduced into yeast function as steroid-dependent transcriptional activators [9–15]. As a result, yeast serve as a useful tool for studying mammalian steroid hormone receptor function in isolation from confounding factors found in mammalian cells. To perform a yeast-based steroid hormone receptor assay, DNA encoding a reporter gene linked to a steroid hormone responsive element is introduced into the yeast along with DNA encoding the appropriate steroid hormone receptor. The stably transfected yeast is incubated with the test chemical for several hours and then lysed and assayed for reporter gene activity. Reporter genes used in these assays produce an easily measurable protein product such as β-galactosidase and luciferase. Chemical interactions with the receptor can be determined by measuring β-galactosidase activity using a standardized enzyme assay on a microplate reader for β-galactosidase or luciferase activity using a luminometer. The ER activity of E2, BPA and HPTE were determined using a yeast-based ER assay (Fig. 2). The results confirm initial findings obtained with a competitive binding assay that, like E2, HPTE and BPA induce ER-mediated reporter gene activity.

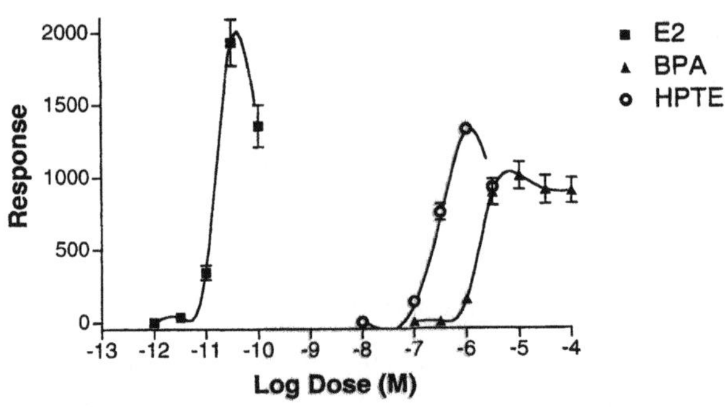

Fig. 2. Yeast-based estrogen receptor assay comparing the relative estrogenic activity of estradiol, HPTE, and bisphenol A. Yeast were incubated overnight with the indicated concentrations. Following incubation the cultures were then assayed for β-galactosidase activity

A yeast-based steroid hormone receptor assay differs from a competitive binding assay in that it not only determines the ability of a chemical to bind to a receptor but also its ability to activate the receptor. Additional advantages of using yeast to study steroid hormone receptor function include ease of manipulation, rapid attainment of stable transformants, ability to process a large number of samples quickly and inexpensively, and limited metabolic capabilities. The yeast-base steroid hormone receptor assay has been used to characterize the interaction of a number of chemicals with the human estrogen, androgen, and progesterone receptors [16–29].

Recently a novel yeast-based assay was described that screened for chemicals with steroid receptor activity by detecting the ability of chemicals to induce interaction of nuclear hormone receptor with coactivators [30]. The assay utilizes a two-hybrid system in which yeast are transformed with nuclear hormone receptors as well as coactivators known to interact with steroid receptors. Transcription is induced if the test chemical interacts with the nuclear receptor and induces a conformation of that receptor that facilitates interaction with the coactivator. This assay may overcome differences in the transcriptional machinery that exists between yeast and mammalian cells.

Yeast assay systems vary in their ability to discriminate between agonists and antagonists. In some systems pure antagonists such as ICI 164,384 behave as agonists [20, 31–34]. Others have reported the ability to detect antagonists using yeast-based assays [24, 26, 35]. A disadvantage of a yeast-based assay is the presence of yeast cell wall and active transport mechanisms that may vary from those found in mammalian cells and may affect the activity and potency of some test chemicals.

4 Mammalian-Based Reporter Gene Assays

Mammalian-based reporter gene assays circumvent some of the problems associated with yeast-based reporter gene assays such as yeast-specific transcriptional factors and differences in cell wall composition. In addition, unlike most yeast-based assays, mammalian-based assays can distinguish between cell-specific steroid receptor agonists and antagonists. A number of different mammalian-based reporter gene assays have been developed to characterize the interaction of chemicals with steroid receptors [3, 4, 6, 36–53]. For some assays, a steroid-responsive reporter gene is transfected into a cell that already contains the steroid receptor of interest, such as an ER-expressing human breast cancer cell line. In other assays, a vector expressing the steroid receptor of interest and a reporter gene are co-transfected into a cell that does not express sufficient levels of the receptor of interest. The reporter gene may be linked either to a synthetic steroid-responsive promoter or to a natural steroid-responsive promoter. Utilization of mammalian cells allows the investigator to determine species, tissue, and promoter-specific steroid hormone receptor-mediated responses to various ligands of interest.

Calcium phosphate, DEAE-dextran, lipofection, and electroporation are the most commonly used transfection methods to introduce (transfect) DNA into cells [54]. The first three methods enhance DNA interactions with the cell surface and therefore augment DNA uptake by endocytosis. Electroporation relies on electrically induced pores through which DNA passively enters. Expression of transfected DNA can be monitored as early as 12 h and as late as 72 h after transfection. Assays performed during this time interval are referred to as transient transfection assays. Transfected genes are gradually lost after 72 h. The amount of DNA taken up by cells during transient transfections often varies significantly. Cotransfection of multiple reporter genes provides a convenient means of controlling for transfection variability. Cells are cotransfected with a DNA mixture composed of two separate plasmids each harboring a different reporter gene, such as a steroid-responsive luciferase reporter gene and a β-galactosidase reporter gene controlled by a constitutively active regulatory element. The resultant activity of the steroid-responsive reporter is normalized to the activity of the constitutively-expressed reporter.

DNA can be introduced into cellular chromatin to form stable transformants at low frequency (0.001 – 1 %). Stable transformants can be isolated under selective pressure. An advantage of a stable transformant is that a heritable genotype is produced. Experiments can be performed without transfecting each time, thereby improving reproducibility and making stable transformants more attractive for major screening efforts. Benefits of transient transfections are that they are more flexible and can be modified to study any receptor or reporter gene of interest. In addition, transient transfections generally have a higher level of reporter gene induction, thus increasing assay sensitivity.

The activities of E2, HPTE, and BPA were examined in human HepG2 hepatoma cells transiently transfected with an estrogen receptor alpha expression plasmid along with an estrogen-responsive luciferase reporter gene and a constitutively expressed β-galactosidase reporter gene (Fig. 3). Transfected cells were incubated overnight in the presence of E2, HPTE, or BPA. Following incubation the cells were lysed and assayed for β-galactosidase and luciferase activ-

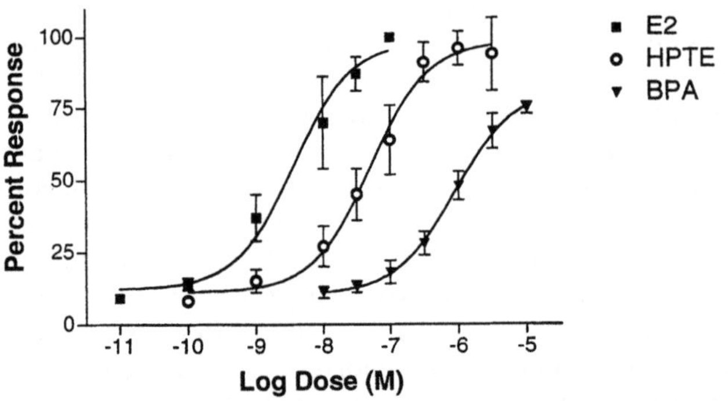

Fig. 3. HepG2 estrogen receptor assay comparing the activity of estradiol, HPTE, and bisphenol A. HepG2 cells were transiently cotransfected with human estrogen receptor, an estrogen responsive luciferase reporter gene, and a constitutively active β-galactosidase reporter gene (transfection control). The transfected cells were treated with estradiol or bisphenol A for 24 h and then assayed for both luciferase and β-galactosidase activities. Results are presented as luciferase values normalized to β-galactosidase activity

ity. The assays were performed in 96-well plates, allowing for rapid quantification of a large number of samples. The potency of BPA and HPTE relative to E2 in the HepG2 cells varies significantly from the relative potency for BPA determined in the yeast assay. The reason for this difference in relative potency between the HepG2 assay and the yeast assay may be due in part to the presence of steroid hormone binding globulin that is in the serum-containing medium used for the HepG2 assay and which is also produced by the HepG2 cells themselves. Some of the E2 added to the HepG2 cultures will bind to steroid hormone binding globulin; as a result, less free hormone will be available for binding to the ER and that will cause a shift in the E2 dose-response curve to the right. This assay has also been used to characterize chemical interactions with the androgen receptor [40].

5
Cell Proliferation Assays

Cell proliferation assays are relatively easy to perform and only require cell culture facilities. Numerous human and rodent steroid-responsive cell lines derived from mammary, uterine, ovarian, and prostatic tissue have been described [38, 55–62].

One of the most commonly used cell lines for assessing chemical interaction with the ER has been the MCF-7 human breast cancer cell line [55, 63, 64]. These cells proliferate in response to estrogen. To perform a cell proliferation assay, MCF-7 cells are plated in a medium containing serum that has been stripped of endogenous hormones. Test chemical is added to the cultures and cells are allowed to proliferate for five to six days. Proliferation can be determined by a number of methods including counting cells or nuclei in a Coulter Counter or by using indices of proliferation including the metabolic reduction of soluble 3-(4,5-dimethylthiazol-2-yl)-2,5-diphenyltetrazolium bromide (MTT) or alamar blue, and incorporation of tritiated thymidine. A chemical is designated an ER agonist if it significantly increases cell proliferation relative to a control. A chemical is an ER antagonist if it inhibits proliferation when added in culture along with an inducing dose of E2. This assay is routinely performed in a 12-well plate but can be adapted to a 96-well format [65].

Cell proliferation assays have several limitations. Other growth regulatory factors and steroids such as insulin, insulin like growth factor, epidermal growth factor, transforming growth factor-β, androgens, and retinoic acid can alter cell proliferation [64, 66–74]. Thus a positive response cannot be attributed strictly to estrogen receptor agonists. In addition, cell lines undergo genetic drift over time and, as a result, multiple strains of these cell lines exist with varying responsiveness to steroids [63, 64, 75]. Thus it is important when establishing a cell proliferation assay to screen for steroid hormone responsiveness. Once a steroid responsive strain has been obtained it must be carefully maintained and monitored.

There are few cell culture models for studying chemical interaction with the androgen receptor. The human prostate cell line LNCaP has been used as a model to study androgen-induced cell proliferation [76–78]. However, the androgen receptor in this cell line has a mutation in the androgen binding domain that alters binding and, as a result, this cell line may not be useful for screening [79]. A nontumorigenic rat prostate epithelial cell line that responds to androgens with enhanced proliferation has been described [60]. Cell lines have been described which show inhibition of proliferation by androgens. MCF-7 cells transfected with the androgen receptor are inhibited from proliferation by androgens [61, 80, 81]. The utility of these cell lines as screening assays for chemical interaction with the androgen receptor has yet to be determined.

6 Peptide Probes

Ligand binding to a steroid hormone receptor causes a conformational change in that receptor. A recent publication described a method that used affinity-selected, phage-displayed peptides to probe the conformational changes that occur with the ER after binding various ligands [82]. In this manuscript it was demonstrated that different peptide-binding surfaces on the ER are exposed in response to binding different ligands.

To perform a phage display assay ER is immobilized on 96-well plates and then incubated for 1 h with test compound (Fig. 4). Phage expressing ER-specific peptides are added directly to the wells and incubated for 30 min. Unbound phage is removed by washing and bound phage is detected by using antibody coupled to horseradish peroxidase (HRP). Assays are developed and absorbance measured in a microplate reader.

The authors demonstrate that a fingerprinting technique using multiple peptides can be used to classify ER ligands quickly into different categories such as agonist (resembling the E2 pattern), antagonist (resembling the ICI 182,780 pattern), mixed agonists/antagonists (resembling the tamoxifen pattern), or novel effectors. Fingerprints may also be used to determine structure activity relationships.

A strength of this fingerprinting technique is that it is broadly applicable to any protein or receptor that undergoes structural changes on binding to a ligand or substrate. This method offers an advantage over a traditional binding assay in that additional information is gained regarding the type of interaction (agonist vs antagonist). Another advantage of this assay over reporter gene assays and cell proliferation assays is that toxicity and metabolism are not a concern. The reliability and utility of this assay for screening structurally diverse endocrine active chemicals remain to be determined.

7 Summary

In summary, our results demonstrate the utility of in vitro assays for characterizing chemical interaction with steroid hormone receptors. These assays are relatively rapid and inexpensive, and they allow for the testing of multiple chemicals over a wide dose range. As a screening tool, in vitro assays cannot account for the complexity of an in vivo response but can help predict the type of re-

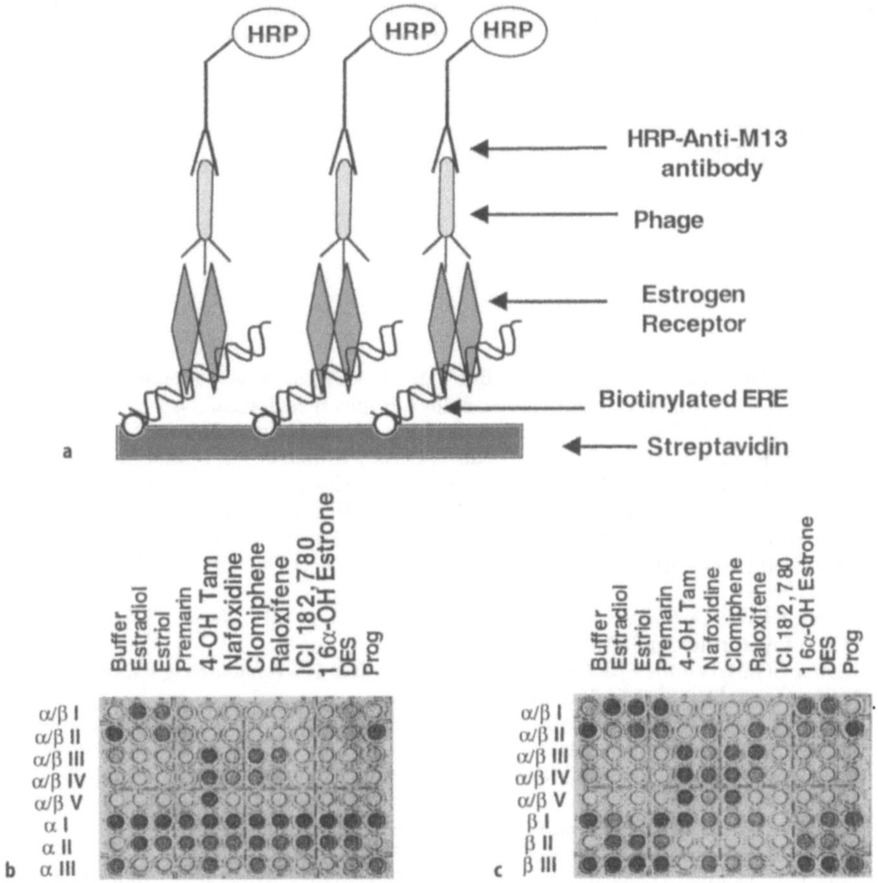

Fig. 4a–c. A Phage display assay. Biotinylated vitellogenin estrogen responsive element (ERE) is immobilized on 96-well plates precoated with streptavidin. The ER is then immobilized on the ERE before the addition of phage. HRP, horseradish peroxidase. **B** Fingerprint analysis of ER modulators on ERα. C. Fingerprint analysis of ER modulators on ERβ. Immobilized ER was incubated with estradiol (1 mmol/l), estriol (1 mmol/l), premarin (10 mmol/l), 4-OH tamoxifen (4-OH Tam, 1 mmol/l), nafoxidine (10 mmol/l), clomiphene (10 mmol/l), raloxifene (1 mmol/l), ICI 182,780 (1 mmol/l), 16α-OH estrone (10 mmol/l), diethylstilbestrol (DES, 1 mmol/l), or progesterone (Prog, 1 mmol/l). Adapted from [82] with permission from the National Academy of Sciences, USA

sponse to expect in vivo. In addition, these assays are useful for gaining mechanistic information that can be used in the design and interpretation of in vivo studies.

Acknowledgements. The financial assistance of the National Institutes of Health (ES04917, ES04917, and DK48807) and the Texas Agricultural Experiment Station is gratefully acknowledged.

8 References

1. Hughes MR, Schrader WT, O'Malley BW (eds). (1991) Laboratory methods manual for hormone action and molecular endocrinology. Houston Biological Associates, Houston
2. Keen M (ed) (1999) Receptor binding techniques. Humana Press, Totowa, NJ
3. Gray LE, Kelce WR, Wiese T, Tyl R, Gaido K, Cook J, Klinefelter G, Desaulniers D, Wilson E, Zacharewski T, Waller C, Foster P, Laskey J, Reel J, Giesy J, Laws S, McLachlan J, Breslin W, Cooper R, DiGiulio R, Johnson R, Purdy R, Mihaich E, Safe S, Sonnenschein C, Welshons W, Miller R, McMaster S, Colborn T (1997) Reprod Toxicol 11:719
4. Shelby MD, Newbold RR, Tully DB, Chae K, Davis VL (1996) Environ Health Perspect 104:1296
5. Kuiper GGJM, Carlsson B, Grandien K, Enmark E, Haggblad J, Nilsson S, Gustafsson JA (1997) Endocrinology 138:863
6. Kuiper GGJM, Lemmen JG, Carlsson B, Corton JC, Safe SH, van der Saag PT, van der Burg P, Gustafsson JA (1998) Endocrinology 139:4252
7. Nakai M, Tabira Y, Asai D, Yakabe Y, Shimyozu T, Noguchi M, Takatsuki M, Shimohigashi Y (1999) Biochem Biophys Res Comm 254:311
8. Bolger R, Wiese TE, Ervin K, Nestich S, Checovich W (1998) Environ Health Perspect 106:551
9. Metzger D, White JH, Chambon P (1988) Nature 334:31
10. Schena M, Yamamoto KR (1988) Science 241:965
11. Mak P, McDonnell DP, Weigel NL, Schrader WT, O'Malley BW (1989) J Biol Chem 264: 21,613
12. McDonnell DP, Nawaz Z, Densmore C, Weigel NL, Pham TA, Clark JH, O'Malley BW (1991) J Steroid Biochem Mol Biol 39:291
13. Wright APH, Carlstedt-Duke J, Gustafsson JA (1990) J Biol Chem 265:14,763
14. Ohashi H, Yang YF, Walfish PG (1991) Biochem Biophys Res Comm 178:1167
15. Purvis IJ, Chotai D, Dykes CW, Lubahn DB, French FS, Wilson EM, Hobden AN (1991) Gene 106:35
16. Jobling S, Reynolds T, White R, Parker MG, Sumpter JP (1995) Environ Health Perspect 103:582
17. Chen CW, Hurd C, Vorojeikina DP, Arnold SF, Notides AC (1997) Biochem Pharmacol 53:1161
18. Coldham NG, Dave M, Sivapathasundaram S, McDonnell DP, Connor C, Sauer MJ (1997) Environ Health Perspect 105:734
19. Collins BM, McLachlan JA, Arnold SF (1997) Steroids 62:365
20. Gaido KW, Leonard LS, Lovell S, Gould JC, Babai D, Portier CJ, McDonnell DP (1997) Toxicol Appl Pharmacol 143:205
21. Klotz DM, Ladlie BL, Vonier PM, McLachlan JA, Arnold SF (1997) Mol Cell Endocrinol 129:63
22. Ramamoorthy K, Vyhlidal C, Wang F, Chen IC, Safe S, McDonnell DP, Leonard LS, Gaido KW (1997) Toxicol Appl Pharmacol 147:93
23. Routledge EJ, Sumpter JP (1997) J Biol Chem 272:3280

24. O'Connor JC, Cook JC, Slone TW, Makovec GT, Frame SR, Davis LG (1998) Toxicol Sci 46:45
25. Odum J, Lefevre PA, Tittensor S, Paton D, Routledge EJ, Beresford NA, Sumpter JP, Ashby J (1997) Regul Toxicol Pharmacol 25:176
26. Sohoni P, Sumpter JP (1998) J Endocrinol 158:327
27. Breinholt V, Larsen JC (1998) Chem Res Toxicol 11:622
28. Breithofer A, Graumann K, Scicchitano MS, Karathanasis SK, Butt TR, Jungbauer A (1998) J Steroid Biochem Mol Biol 67:421
29. Graumann K, Breithofer A, Jungbauer A (1999) Sci Total Environ 225:69
30. Nishikawa J, Saito K, Goto J, Dakeyama F, Matsuo M, Nishihara T (1999) Toxicol Appl Pharmacol 154:76
31. Metzger D, Losson R, Bornert JM, Lemoine Y, Chambon P (1992) Nucleic Acids Res 20:2813
32. Pham TA, Hwung YP, Santiso MD, McDonnell DP, O'Malley BW (1992) Mol Endocrinol 6:1043
33. Wrenn CK, Katzenellenbogen BS (1993) J Biol Chem 268:24,089
34. Kohno H, Gandidi O, Curtis SW, Korach KS (1994) Steroids 1994:572
35. Routledge EJ, Parker J, Odum J, Ashby J, Sumpter JP (1998) Toxicol Appl Pharmacol 153:12
36. Garner CE, Jefferson WN, Burka LT, Matthews HB, Newbold RR (1999) Toxicol Appl Pharmacol 154:188
37. Vinggaard AM, Bonefeld Joergensen EC, Larsen JC (1999) Toxicol Appl Pharmacol 155:150
38. CastroRivera E, Safe S (1998) J Steroid Biochem Mol Biol 64:287
39. Gould JC, Leonard LS, Maness SC, Wagner BL, Conner K, Zacharewski T, Safe S, McDonnell DP, Gaido KW (1998) Mol Cell Endocrinol 142:203
40. Maness SC, McDonnell DP, Gaido KW (1998) Toxicol Appl Pharmacol 151:135
41. Pennie WD, Aldridge TC, Brooks AN (1998) J Endocrinol 158:R11
42. Ram PT, Kiefer T, Silverman M, Song Y, Brown GM, Hill SM (1998) Mol Cell Endocrinol 141:53
43. Ruh MF, Bi Y, Cox L, Berk D, Howlett AC, Bellone CJ (1998) Endocrine 9:207
44. Snoek R, Bruchovsky N, Kasper S, Matusik RJ, Gleave M, Sato N, Mawji NR, Rennie PS (1998) Prostate 36:256
45. Zacharewski TR, Meek MD, Clemons JH, Wu ZF, Fielden MR, Matthews JB (1998) Toxicol Sci 46:282
46. Beck S, Fegert P, Gott P (1997) Int J Oncol 10:1051
47. Edmunds J, Fairey ER, Ramsdell JS (1997) Neurotoxicology 18:525
48. Jorgensen E, Autrup H, Hansen JC (1997) Carcinogenesis 18:1651
49. Kramer VJ, Helferich WG, Bergman A, KlassonWehler E, Giesy JP (1997) Toxicol Appl Pharmacol 144:363
50. Ramamoorthy K, Wang F, Chen IC, Norris JD, McDonnell DP, Leonard LS, Gaido KW, Bocchinfuso WP, Korach KS, Safe S (1997) Endocrinology 138:1520
51. Gao T, Marcelli M, McPhaul MJ (1996) J Steroid Biochem Mol Biol 59:9
52. Steinmetz R, Young P, Caperellgrant A, Gize EA, Madhukar BV, Benjonathan N, Bigsby RM (1996) Cancer Res 56:5403
53. Kelce WR, Stone CR, Laws SC, Gray LE, Kemppainen JA, Wilson EM (1995) Nature 375:581
54. Alam J, Cook JL (1990) Anal Biochem 188:245
55. Soto AM, Lin T-M, Justicia H, Silvia RM, Sonnenschein C (1992) An "in culture" bioassay to assess the estrogenicity of xenobiotics (E-screen). In: Colburn T, Clement C (eds) Chemically induced alterations in sexual and functional development: the wildlife/human connection. Princeton Scientific Publishing, Princeton, NJ, p 295
56. Soto AM, Michaelson CL, Prechtl NV, Weill BC, Sonnenschein C, Olea-Serrano F, Olea N (1998) Adv Exp Med Biol 444:9
57. Tabibzadeh S, Kaffka KL, Kilian PL, Satyaswaroop PG (1990) In Vitro Cell Dev Biol 26:1173

58. Baldwin WS, Curtis SW, Cauthen CA, Risinger JI, Korach KS, Barrett JC (1998) In Vitro Cell Dev Biol Animal 34:649
59. Holinka CF, Hata H, Kuramoto H, Gurpide E (1986) J Steroid Biochem 24:85
60. Lucia MS, Sporn MB, Roberts AB, Stewart LV, Danielpour D (1998) J Cell Physiol 175:184
61. Janssen T, Raviv G, Camby I, Petein M, Darro F, Pasteels JL, Schulman C, Kiss R (1995) Int J Oncol 7:1219
62. Langdon SP, Ritchie A, Young K, Crew AJ, Sweeting V, Bramley T, Hillier S, Hawkins RA, Tesdale AL, Smyth JF, Miller WR (1993) Int J Cancer 55:459
63. Wiese TE, Kral LG, Dennis KE, Butler WB, Brooks SC (1992) In Vitro Cell Dev Biol 28A:595
64. Desaulniers D, Leingartner K, Zacharewski T, Foster WG (1998) Toxicol In Vitro 12:409
65. Korner W, Hanf V, Schuller W, Kempter C, Metzger J, Hagenmaier H (1999) Sci Total Environ 225:33
66. Burak WE, Quinn AL, Farrar WB, Brueggemeier RW (1997) Breast Cancer Res Treat 44:57
67. Durgam VR, Fernandes G (1997) Cancer Lett 116:121
68. Nodland KI, Wormke M, Safe S (1997) Arch Biochem Biophys 338:67
69. Panno ML, Salerno M, Pezzi V, Sisci D, Maggiolini M, Mauro L, Morrone EG, Ando S (1996) J Cancer Res Clin Oncol 122:745
70. Venturelli E, Coradini D, Gornati D, Secreto G (1996) Int J Oncol 8:687
71. Mazars P, Barboule N, Baldin V, Vidal S, Ducommun B, Valette A (1995) FEBS Lett 362:295
72. Taylor JA, Grady LH, Engler KS, Welshons WV (1995) Breast Cancer Res Treat 34:265
73. Rubin M, Fenig E, Rosenauer A, Menendezbotet C, Achkar C, Bentel JM, Yahalom J, Mendelsohn J, Miller WH (1994) Cancer Res 54:6549
74. De Leon DD, Wilson DM, Powers M, Rosenfeld RG (1992) Growth Factors 6:327
75. Villalobos M, Olea N, Brotons JA, Oleaserrano MF, Dealmodovar J, Pedraza V (1995) Environ Health Perspect 103:844
76. Soto AM, Lin TM, Sakabe K, Olea N, Damassa DA, Sonnenschein C (1995) Oncol Res 7:545
77. Lu S, Tsai SY, Tsai MJ (1997) Cancer Res 57:4511
78. Okamoto M, Lee C, Oyasu R (1997) Endocrinology 138:5071
79. Veldscholte J, Ris-Stalpers C, Kuiper GG, Jenster G, Berrevoets C, Claassen E, van Rooij HC, Trapman J, Brinkmann AO, Mulder E (1990) Biochem Biophys Res Commun 173:534
80. Hackenberg R, Schulz KD (1996) J Steroid Biochem Mol Biol 56:113
81. Szelei J, Jimenez J, Soto AM, Luizzi MF, Sonnenschein C (1997) Endocrinology 138:1406
82. Paige LA, Christensen DJ, Gron H, Norris JD, Gottlin EB, Padilla KM, Chang C, Ballas LM, Hamilton PT, McDonnell DP, Fowlkes DM (1999) Proc Natl Acad Sci USA 96:3999

CHAPTER 3

Antiandrogenic Effects of Environmental Endocrine Disruptors

William R. Kelce[1,2], Elizabeth M. Wilson[2,3]

[1] Metabolism and Safety Evaluation, Searle, K332 A Mail Code 1825, 4901 Searle Parkway, Skokie, IL 60077, USA
e-mail: william.v.kelce@pharmacia.com

[2] Laboratories for Reproductive Biology and the Departments of Pediatrics

[3] Biochemistry and Biophysics, University of North Carolina, Chapel Hill NC 27599-7500, USA
e-mail: emw@med.unc.edu

Steroid hormone receptors regulate embryonic development and sex differentiation by acting as ligand inducible transcription factors. Disrupting these processes can result in transient yet irreversible developmental defects. Several environmental chemicals have recently been identified with antiandrogen activity that have the potential to disrupt normal male sex development in utero. Detection of chemicals with the potential to disrupt normal androgen action is critical to protect human and ecological health. The molecular mechanisms by which several of these environmental chemicals act have been characterized and yielded insight into the development of in vitro and in vivo screening assays to detect chemicals with antiandrogen activity. These mechanisms and efforts to develop screening assays are discussed in the context of current endocrine disruptor screening and testing strategies.

Keywords. Antiandrogens, Endocrine disruptor, Sex differentiation and development, Androgen action, Antiandrogen screening assays

1 Introduction and Background 40

2 Mechanisms of Androgen Action 43

3 Alterations in Sex Differentiation 45

4 Environmental Antiandrogens 46
4.1 Dicarboximide Fungicides 47
4.2 *p,p'*-DDT 48
4.3 Methoxychlor 48

5 Additivity Effects of Environmental Antiandrogens: Mixtures and Common Mechanisms 49

6 Screening Assays for Measuring AR Agonist and Antagonist Activities of Environmental Chemicals 54
6.1 Ligand Binding 54
6.2 DNA Binding and Transcriptional Activation Assays 55
6.3 N/C Interaction Assay 55
6.4 In Vivo Assays 58

The Handbook of Environmental Chemistry Vol. 3, Part L
Endocrine Disruptors, Part I
(ed. by M. Metzler)

7 **Conclusions** 59

8 **References** 59

List of Abbreviations

AR	androgen receptor
DDT	1,1,1-trichloro-2,2-bis(*p*-chlorophenyl)ethane
DHT	5α-dihydrotestosterone
ED	endocrine disruptor
EDSP	EPA Endocrine Disruptor Screening Program
EDSTAC	Endocrine Disruptor Screening and Testing Advisory Committee
EPA	US Environmental Protection Agency
HPTE	2,2-bis(*p*-hydroxyphenyl)-1,1,1-trichloroethane
M1	{2-[(3,5-dichlorophenyl)-carbamoyl]oxy}-2-methyl-3-butenoic acid
M2	3,5-dichloro-2-hydroxy-2-methylbut-3-enanilide
methoxychlor	2,2-bis(*p*-methoxyphenyl)-1,1,1-trichloroethane
MPA	medroxyprogesterone acetate
N/C	AR NH_2-/carboxyl-terminal
OECD	Office of Economic Cooperation and Development
p,p'-DDE	1,1-dichloro-2,2-bis(*p*-chlorophenyl)ethylene
procymidone	*N*-(3,5-dichlorophenyl)-1,2-dimethylcyclopropane-1,2-dicarboximide
R1881	methyltrienolone
US	United States
vinclozolin	3-(3,5-dichlorophenyl)-5-methyl-5-vinyloxazolidine-2,4-dione

1 Introduction and Background

Chemicals enter the environment through a number of routes, the most prevalent being from the use of pesticides, herbicides, and fungicides and from waste water. Some of these chemicals have structural features that allow them to bind to different classes of nuclear receptors within the endocrine system.

Steroid hormone receptors regulate embryonic development and sex differentiation by acting as ligand inducible transcription factors. Disrupting these processes can result in transient yet irreversible developmental defects. Several environmental chemicals have recently been identified with antiandrogen activity that have the potential to disrupt normal male sexual development in utero. In all instances thus far, the parent chemicals that were used in the environment as fungicides or pesticides themselves were mostly biologically inactive in terms of endocrine mechanisms. Each had to be metabolized in the body or bioactivated in the environment to compounds with potent antiandrogen activity. The environmental antiandrogens discovered thus far include the M2 metabolite of vinclozolin, a fungicide used widely in the treatment of fruit crops. A second is *p,p'*-DDE which is the long acting stable metabolite of *p,p*-DDT, a pesticide that was banned in the United States but continues to be used in neighboring countries such as Mexico from where the US imports food products. The most recent environmental antiandrogen identified is HPTE, the active metabolite of methoxychlor and an ingredient presently in use in household pesticides and marketed throughout the United States.

The study of these so-called environmental endocrine disruptors (EDs) is critically important at this time because a wide variety of chemicals have the potential to impact adversely human reproductive health. Adverse reproductive effects are being observed in wildlife populations living in highly contaminated areas, such as the demasculinization of male alligators in Lake Apopka, Florida [1]. Inadvertent exposure to environmental chemicals may account for the reported increases in human developmental abnormalities such as hypospadias in the newborn male. The above-mentioned demasculinization of male alligators was due to contamination with *p,p*-DDT and other organochlorine pesticides. Our studies have focused on environmental chemicals that cause developmental reproductive effects through mechanisms directly involving the androgen receptor (AR). Other pathways for endocrine disruption include the inhibition of steroid hormone biosynthesis, transport, and degradation.

Public concern with EDs has driven the United States public policy makers to mandate, via the Food Quality Protection Act (Public Law 104–170) and the Safe Drinking Water Act (Public Law 104–182), that the US Environmental Protection Agency (EPA) develop and implement a screening and testing program for environmental chemicals that produce estrogen-like, antiandrogen, or other effects as deemed appropriate by the EPA. While media attention and scientific reports have been instrumental in bringing the field of endocrine disruption to the forefront of scientific research, definitive data will be required to ensure public safety.

As scientists in this developing field of endocrine toxicology, we believe that the mechanistic information inherent to in vitro screening assays should be combined with information from more apical short-term in vivo tests in order to support hazard assessment issues for endocrine active chemicals. The development of specific ED screening strategies will improve our ability to identify and evaluate, in the context of production volume and environmental exposure data, the potential of chemicals to induce adverse effects in wildlife and/or human populations. In addition to efforts within the United States, other countries led primarily by Japan, and the Office of Economic Cooperation and Development (OECD) member countries, are developing approaches to screen chemicals for endocrine activity. Standardization of definitive, sensitive and accurate ED screening assays among all countries should be a high priority.

In response to the US Congressional mandates outlined above, the EPA formed the Endocrine Disruptor Screening and Testing Advisory Committee (EDSTAC). The EDSTAC committee was comprised of stakeholders from government, industry, academia, and special interest groups and was charged with recommending a strategy to screen chemicals for endocrine activity. The EDSTAC recommended that EPA broaden its scope to include screening chemicals for estrogen, androgen, and thyroid hormone activity, and to expand the screening process itself to include non-mammalian species such as avian, amphibian, and aquatic species. The comprehensive screening program recommended by EDSTAC is a two-tiered strategy. Tier I screening assays are designed to detect chemicals with the ability to interact with the endocrine system, while tier II testing determines and characterizes the endocrine disrupting effects for subsequent hazard assessment.

Before implementation of tier I screening, EDSTAC recommended that the 87,000 chemicals in the EPA database be prioritized using factors such as production volume, potential for human exposure, and high-throughput prescreening for estrogen, androgen, and thyroid hormone receptor activity. While industry in the United States generally applauds a prioritization process, there has been concern that high-throughput prescreening could prematurely and perhaps inappropriately label a commodity chemical as an ED. A concern among industry is the availability of definitive tier I screens at the time of high-throughput prescreening so that positive chemicals can be definitively assessed in more comprehensive tier I screens before being prematurely labeled and deselected by the public. The role and implementation of high-throughput prescreening in the EPA Endocrine Disruptor Screening Program (EDSP) has yet to be defined.

Assays recommended for tier I screening include in vitro receptor binding and transcriptional activation assays together with short-term in vivo assays. The objective of tier I screening is to determine the ability of chemicals to interact with the estrogen or androgen receptor and/or influence thyroid hormone action. Transcriptional activation assays are preferred over receptor binding because agonist and antagonist activities can be distinguished. The ability of increasing concentrations of exogenous chemical to induce reporter gene expression in a dose-response manner is indicative of agonist activity, while the ability of the same doses to reduce agonist-stimulated expression is indicative of antagonist activity. Antagonist activity must be distinguished from cell cytotoxicity, as both mechanisms reduce reporter gene expression.

Assays recommended for in vivo tier I screening include short term animal studies such as the Hershberger assay. This assay assesses the ability of exogenous chemicals to increase sex accessory organ wet/dry weight in either immature intact male rats or castrated adult male rats. Its basis is that androgens bind the AR in sex accessory organs to increase the rate of cell growth and/or decrease the rate of apoptosis. The Hershberger assay has been a standard bioassay for antiandrogen activity in the pharmaceutical industry for years. Both the immature and adult rat assays detect ED activity; however, the timing of dosing is far more critical with intact immature rats, as the results can be confounded by changes in body weight. Both assays are being examined in detail in the United States, Europe, and Japan. In addition to the Hershberger assay, a 15-day adult or 28-day pubertal male repeat-dosing assay has been recommended to screen chemicals for antiandrogen as well as thyroid hormone activity. Details of these procedures can be found in appendix L in the finalized EDSTAC document on the world wide web at http://www.epa.gov/opptintr/opptendo/. Concerted efforts to develop and standardize these in vitro and in vivo test methods among countries and regulatory agencies is an important goal. We favor accreditation of endocrine disruptor testing labs based on performance in assessing ED activity using a standard general protocol and a set of standard chemicals with known activity. Predefined assay performance criteria would then be used to assess the ability of the individual labs and their specific protocols to conduct adequately ED studies.

This review focuses on the molecular mechanisms by which environmental antiandrogens inhibit androgen activity using as examples specific environmental antiandrogens discovered to date. Androgen action and the role of androgens in normal sex differentiation and development are discussed. Data indicating additive interactions at the level of the AR are presented as examples of mixtures acting through common mechanisms, a topic important to the characterization of relative risk to human and ecological health. Finally, suggestions are presented to harmonize tier I in vivo screening assays among the EPA Endocrine Disruptor Screening Program, the High-Production Volume screening program, and the Children's Health Test Rule. The reader is referred elsewhere for discussions of general mechanisms of environmental EDs [2–5], effects of environmental EDs on wildlife [6, 7], effects of environmental estrogens in the male [8], clinical implications of these chemicals in humans [5, 9], and proposed research needs for risk assessment [10].

2 Mechanisms of Androgen Action

Testosterone is the major circulating form of the biologically active androgens. Testosterone is metabolized to 5α-dihydrotestosterone (DHT) (Fig. 1) in peripheral tissues. Both testosterone and DHT bind the same AR; however, DHT is more potent because of its slower dissociation rate [11] which imparts greater stabilization of the AR against degradation [12]. High affinity androgen binding induces an intermolecular interaction between the NH_2- and carboxyl-terminal regions of AR that facilitates AR dimerization and stabilization [13, 14]. The N/C mediated dimerization is required for agonist activity at low physiological androgen concentrations [15]. Receptor dimerization is necessary for specific DNA binding and subsequent transcriptional activation [16].

Antiandrogen binding induces a different conformation of the AR which can disrupt the NH_2-/carboxyl terminal interaction. In most cases of moderate affinity antagonists, this results in loss of DNA binding and thus loss of androgen induced transcriptional activation. However, certain higher affinity antiandrogens may compete for androgen binding but display partial agonist activity at sufficiently high concentrations. It remains to be established by what mechanisms higher affinity antagonists have dual functions as androgen antagonists and agonists. Thus far androgen antagonists, particularly those identified among the environmental chemicals, are amongst a group known as type 1 antagonists. These bind steroid receptors but prevent receptor binding to DNA. This applies to the metabolites of the fungicide, vinclozolin [17], and the pesticide, *p,p'*-DDT [18]. There are several possible mechanisms for AR inhibition by low to moderate affinity binding chemicals. These include increased AR degradation due to the absence of the NH_2-/carboxyl-terminal interaction or in-

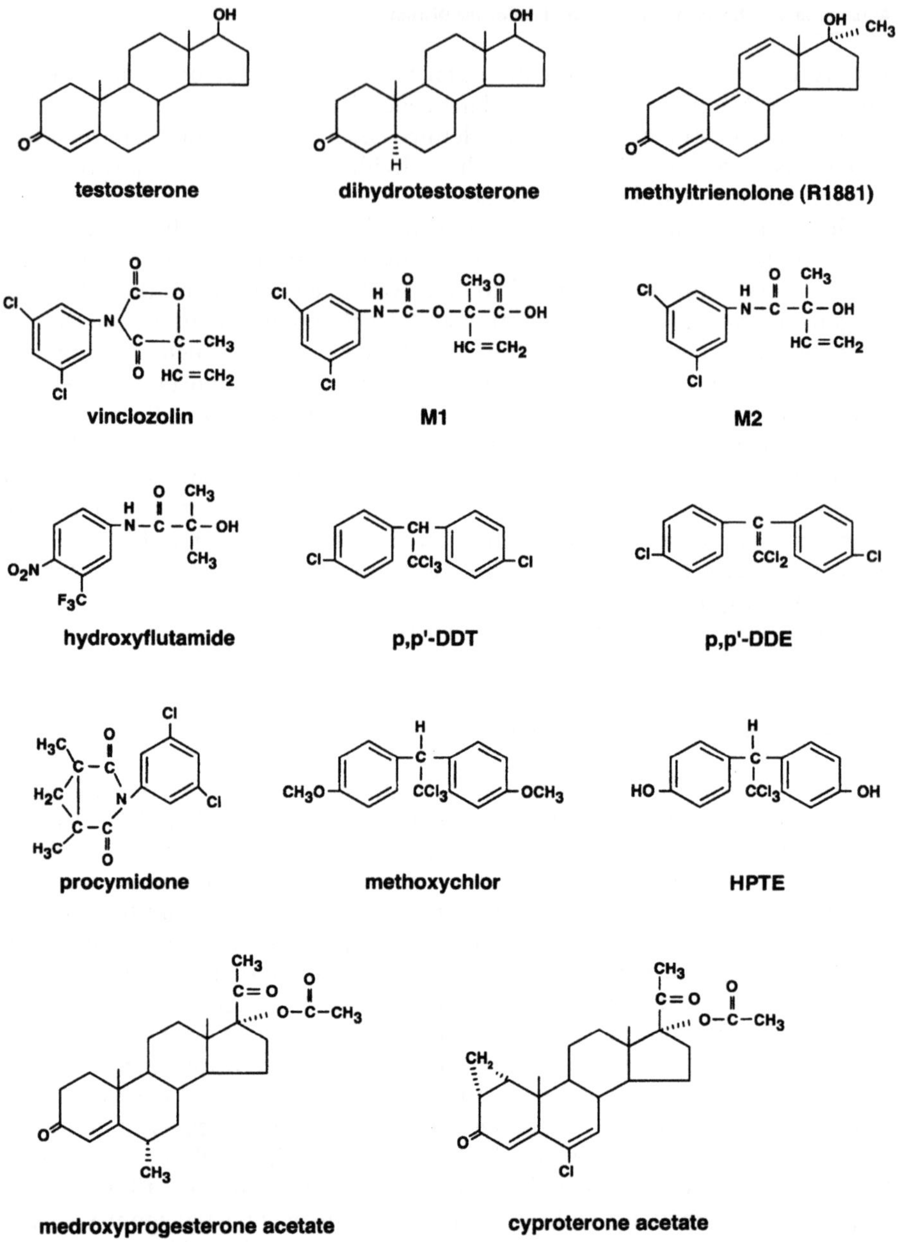

Fig. 1. Chemical formulae of androgens, antiandrogens, and environmental antiandrogens and parent compounds. Shown are the chemical structures of the biologically active androgens, testosterone and dihydrotestosterone; the synthetic androgen methyltrienolone (R1881); the fungicide vinclozolin (3-(3,5-dichlorophenyl)-5-methyl-5-vinyloxazolidine-2,4-dione) and its metabolites, M1 ({2-[(3,5-dichlorophenyl)-carbamoyl]oxy}-2-methyl-3-butenoic acid) and M2 (3,5-dichloro-2-hydroxy-2-methylbut-3-enanilide); the antiandrogens hydroxyflutamide and cyproterone acetate; the pesticide *p,p'*-DDT and its biologically stable metabolite, *p,p'*-DDE; the fungicide procymidone; the pesticide methoxychlor (2,2-bis(*p*-methoxyphenyl)-1,1,1-trichloroethane) and its metabolite HPTE (2,2-bis(*p*-hydroxyphenyl)-1,1,1-trichloroethane); and the synthetic progestin medroxyprogesterone acetate (Provera)

creased dissociation of bound ligands that fail to stabilize the AR. Environmental antiandrogens may prevent AR dimerization or cause the release of receptor associated proteins that are necessary for specific DNA binding and transcriptional activity, or for the association of corepressors that inhibit transcriptional activation.

3
Alterations in Sex Differentiation

Disruption of male sex differentiation and development in utero is not lethal. Deviations in sex development of the newborn have been observed in wildlife and humans populations. Genetic sex determined at fertilization results in the expression of the SRY gene on the Y chromosome [19] which directs differentiation of the indifferent gonad into testes that produce testosterone, which in turn induces differentiation of the male internal ducts and external genitalia [20]. During human fetal development, testosterone synthesis begins at about 65 days gestation, inducing the formation of the epididymides, vas deferens, and seminal vesicles from the Wolffian ducts. DHT, a testosterone metabolite, induces development of the prostate and male external genitalia. In the absence of testosterone, or in complete androgen blockade due to antiandrogens or to mutations in the AR gene, the female external phenotype is expressed in the 46XY individual independent of the presence of an ovary. A single gene for the AR is expressed on the long arm of the X chromosome at Xq11–12 [21] that codes for the 114 kDa AR comprised of 919 amino acids [22].

An important consideration in establishing the potential detrimental effects of environmental EDs is their ability to interfere with fetal development, particularly in the male. Since the male fetus depends on androgen exposure at critical periods during gestation, competing chemicals at this time have the potential to cause irreversible developmental changes as the male reproductive system is sensitive to low dose chemical disruption. Even though chemical exposure may be transient, the effects on reproductive capacity can be irreversible. The enhanced sensitivity of the developing male fetus to antiandrogen effects of environmental chemicals likely results from reduced levels of circulating endogenous androgens. Thus, endocrine disruption in the fetus is more likely to occur than in the adult environment where circulating androgen levels are higher. Another consideration that complicates cause and effect evaluations of environmental chemicals is that functional alterations may not manifest until later in life. It is therefore the developing fetus that is a primary focus for endocrine disruption by environmental chemicals with antiandrogen activity. Our studies have focused on environmental chemicals that cause developmental reproductive effects through mechanisms directly involving the AR. Other pathways for endocrine disruption include the inhibition of steroid hormone biosynthesis, transport, and degradation.

4 Environmental Antiandrogens

The properties of several environmental chemicals that have antiandrogen activity both in vitro and in vivo are described below. To date relatively few chemicals have been tested for ED activity. This results in part from the labor intensive methods required for detecting these activities. However, as in vitro methods for screening and testing chemicals for ED activity become automated, more will likely be identified. Furthermore, more diverse mechanisms of inhibition may be discovered that involve receptor stabilization, DNA binding, and coactivator interactions. A summary of the properties of the environmental antiandrogens discovered thus far is shown in Table 1.

The environmental antiandrogens identified have only moderate to low binding affinity for the AR in the micromolar range, compete for testosterone or DHT binding to the AR, and inhibit AR binding to androgen response element DNA resulting in inhibition of transcriptional activation [17]. However, some environmental compounds may be discovered that belong to a group of higher affinity antagonists that display partial agonist activity at high concentrations. It is the dilemma of both antagonist and agonist activity by the same compound that has raised mechanistic questions.

We [17] and others [23, 24] have proposed that antagonism of steroid receptors results from the formation of dimers where each steroid receptor monomer is bound with a different ligand. Same ligands bound to both receptor monomers would result in agonism. These mechanisms could explain why, under some conditions, antagonists are partial agonists as recently shown for estriol. When administered repeatedly, estriol is an agonist, but when given in the presence of estradiol it acts as an antagonist [24]. We observed similar results with pharmaceutical antagonists, although as yet not with the environmental antiandrogens.

Table 1. Summary of the properties of three environmental chemicals with antiandrogen activity. The pesticides DDT and methoxychlor are each metabolized to active compounds *p,p'*-DDE and HPTE, respectively, which bind the AR with moderate affinity and act to inhibit transcriptional activation mediated by the AR [18, 37]. Properties of the active metabolite M2 of the fungicide vinclozolin were previously described [17, 27]. Inhibition constants (K_I) were determined using rat ventral prostate cytosol as previously described [18]. The N/C interaction, DNA binding and transactivation assays were performed as previously described [13–15, 17]

Environmental antiandrogens

Environmental use	Parent compound	Active metabolite	K_I µmol/l	Inhibit AR N/C interaction	Inhibit AR DNA binding/transact
Pesticide	DDT	*p,p'*-DDE	3.5	+	+
Pesticide	Methoxychlor	HPTE	1.4	+	+
Fungicide	Vinclozolin	M2	9.7	+	+

+ indicates activity detected in the assay.

It is conceivable that the formation of mixed-ligand dimers causes rapid dissociation of androgen. Under normal conditions, androgen dissociates from the AR with slow kinetics, with a dissociation $t_{1/2}$ of about 2.8 h at 37 °C determined using the synthetic radiolabeled, high-affinity AR agonist, [^{3}H]methyltrienolone ([^{3}H]R1881) [12]. Preliminary studies with certain high affinity antagonists indicate that they dissociate from AR with rapid kinetics, $t_{1/2}$ 5 min at 37 °C [15]. The presence of an antagonist in the AR dimer may enhance the dissociation of bound androgen. AR would be more rapidly degraded and thus unavailable to activate transcription, resulting in androgen antagonism. This putative mechanism may account for the antiandrogen activity of the environmental chemicals discovered thus far. The mixed-ligand dimer-induced rapid dissociation is consistent with empirical data for the dicarboximide fungicides, *p,p'*-DDE and methoxychlor.

4.1 Dicarboximide Fungicides

Vinclozolin (3-(3,5-dichlorophenyl)-5-methyl-5-vinyloxazolidine-2,4-dione) (see Fig. 1) is a dicarboximide fungicide used in the United States and Europe to protect crops from fungal disease. In studies performed at the US EPA, exposure of pregnant rats to 100 mg/kg/day vinclozolin from gestational day 14 to postnatal day 3 caused reduced anogenital distance in male pups, nipple development, cleft phallus with hypospadias, suprainguinal ectopic testes, vaginal pouches, and atrophy of the seminal vesicles and prostate [25, 26]. Exposure to 50 mg/kg/day vinclozolin during gestational day 14 to postnatal day 3, the critical period of sex differentiation in the rat, resulted in infertility and reduced sperm counts in adult male offspring associated with severe hypospadias. The results confirm that the developing fetus is sensitive to endocrine disruption by environmental chemicals.

The antiandrogen activity of vinclozolin results from its metabolic conversion to M1 ({2-[(3,5-dichlorophenyl)-carbamoyl]oxy}-2-methyl-3-butenoic acid) and M2 (3,5-dichloro-2-hydroxy-2-methylbut-3-enanilide) (Fig. 1). It is these biologically active metabolites that inhibit androgen binding to AR [27] with subsequent inhibition of AR DNA binding and transcriptional activation [4]. It is noteworthy that there is structural similarity between M1, M2, and the potent antiandrogen, hydroxyflutamide [17]. Molecular studies with transfected CV1 cells expressing human AR indicate that both M1 and M2 inhibit androgen-induced transcriptional activation. Electrophoretic mobility DNA band shift assays demonstrate an inability of antiandrogen-bound AR to bind androgen response element DNA, accounting for transcriptional inhibition [17].

Procymidone (*N*-(3,5-dichlorophenyl)-1,2-dimethylcyclopropane-1,2-dicarboximide) (Fig. 1) is another dicarboximide fungicide currently in use in Canada, Europe, and Japan, and approval is being sought in the United States. The parent compound has a low affinity for AR [28, 29] but preliminary evidence indicates that treatment of COS cells expressing human AR with procymidone causes AR import to the nucleus and inhibition of AR androgen binding. In transcriptional activation studies using CV-1 cells, procymidone inhibits DHT-induced transactivation [30]. Procymidone has been shown in earlier studies at the Environmental Protection Agency to inhibit male sexual differentiation following in utero exposure similar to the results with vinclozolin, although the potency of procymidone was approximately 2.5-fold less than that of vinclozolin [30]. When administered to rats, procymidone causes increased serum testosterone levels, and in competitive binding studies has a low affinity for AR [28, 29]. Like the other environmental antiandrogens, procymidone is likely metabolized to an active compound yet to be identified.

4.2 *p,p'*-DDT

DDT (1,1,1-trichloro-2,2-bis(*p*-chlorophenyl)ethane) is an organochlorine pesticide that was banned from use in the United States in 1973. Yet DDT continues to be produced and shipped to neighboring countries and worldwide. DDT is metabolized to *p,p'*-DDE (1,1-dichloro-2,2-bis(*p*-chlorophenyl)ethylene) (Fig. 1) which bioaccumulates in the environment with a half-life in the body of more than 50 years. When administered to pregnant rats from gestational day 14 to 18, *p,p'*-DDE (100 mg/kg/day) reduces anogenital distance and causes retention of thoracic nipples in male progeny [18], both indicative of prenatal antiandrogen exposure [31]. *p,p'*-DDE binds AR with moderate affinity and inhibits DHT-induced transcriptional activation with a potency similar to that of the antiandrogen, hydroxyflutamide [18]. Although DDT persists in the environment, it remains to be established to what extent it is a threat to the human population, particularly from imported food. Its effectiveness in malaria control made it the pesticide of choice in many developing countries.

The in vivo and in vitro effects of environmental antiandrogens were recently summarized [10].

4.3 Methoxychlor

Methoxychlor (2,2-bis (*p*-methoxyphenyl)-1,1,1-trichloroethane) is a DDT analogue currently in use as a pesticide in the United States and worldwide. Like the environmental antiandrogens described above, methoxychlor is metabolized to an active metabolite, HPTE (2,2-bis(*p*-hydroxyphenyl)-1,1,1-trichloroethane) (Fig. 1). Methoxychlor is a well recognized environmental estrogen that increases uterine weight, glycogen deposition, and ornithine decarboxylase activity in immature or ovariectomized adult rat uterus [32–34] and causes precocious vaginal opening, vaginal cornification, and permanent estrus in adult female mice and rats exposed neonatally [35, 36]. Methoxychlor induces estrogen receptor dependent MCF-7 cell proliferation and reporter gene activity following transient and stable transfection [37]. Despite its known estrogen activity, methoxychlor and HPTE were tested for antiandrogenic activity. In addition to confirming the estrogenic activity of HPTE using a human mammary carcinoma MCF-7 cell proliferation assay and stable transfection assays, HPTE is a potent environmental antiandrogen [37]. Methoxychlor administered to adult male rats delays the age of balano-preputial separation and decreases seminal vesicle weight despite normal plasma levels of testosterone, prolactin, and luteinizing hormone, suggesting a dual potential to act as an estrogen agonist and androgen antagonist. HPTE is likely the active antiandrogen because it, and not the parent compound, methoxychlor, competes for AR binding of [^{3}H]R1881 and inhibits androgen-induced transcriptional activation in transient transfection assays. Like the metabolites of DDT and vinclozolin, HPTE acts as an antiandrogen by inhibiting AR binding to androgen response element DNA [37]. In light of the apparent dual activities of HPTE as an ER agonist and AR antagonist in cell-free equilibrium binding assays, transfected cell lines, and in vivo, HPTE has the potential to disrupt reproductive development and function through separate steroid receptor pathways. As some male developmental processes are adversely affected by both ER agonists and AR antagonists, these results provide a mechanism for synergistic toxicity by single environmental chemicals through dual steroid receptor pathways.

5 Additivity Effects of Environmental Antiandrogens: Mixtures and Common Mechanisms

Single chemical exposures are rare in biological systems. Even in the laboratory setting where test systems such as cells, tissue slices, or animals are dosed with presumed single chemicals, exposure is often to a complex mixture of vehicle, test chemical metabolites, and/or stereoisomers of the parent chemical. In the field or work environment, multiple chemical exposures are commonplace. The issue of multiple chemical exposure and chemical interactions has surfaced in the ED field. A question of fundamental importance to the US EPA is how to assess the effects of exposure to mixtures of endocrine active environmental chemicals. This task is complicated because single formulated products purchased by the consumer are often themselves mixtures of surfactants, safeners, and active ingredients, each of which may exhibit hormone or antihormone activity. In the simplest case of chemical mixtures acting through a common receptor mechanism, empirical data suggest that interaction effects of antiandrogens on AR-induced transcriptional activation are approximately additive in magnitude rather than synergistic, suggesting that a toxic equivalency factor approach is appropriate.

The toxic equivalency factor method was proposed for use in human and ecological risk assessment for polyhalogenated aromatic hydrocarbons acting at the level of the arylhydrocarbon receptor [38]. This method assigns a fractional potency of a chemical relative to a standard and is calculated as the ratio of ED50 (the dose that causes 50% effect) values. Fractional potency is multiplied by the tissue concentration of the chemical to determine its toxic equivalence relative to the standard. The toxic equivalency factor method assumes that chemical interaction is additive, so that toxic equivalents for all chemicals in a complex mixture can be simply added to determine the total equivalents concentration. The assumption of additivity and the toxic equivalency factor approach appears to be appropriate for risk assessment of chemicals acting

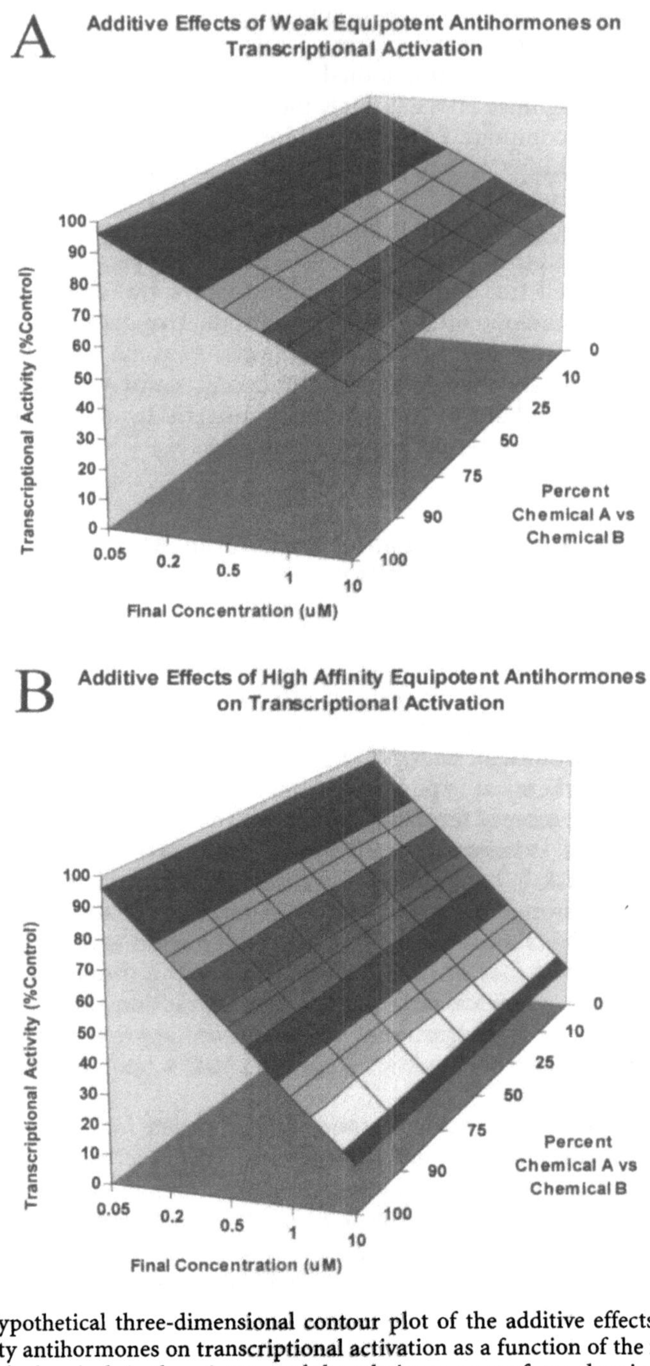

Fig. 2 A, B. Hypothetical three-dimensional contour plot of the additive effects of: A weak; B high affinity antihormones on transcriptional activation as a function of the final concentration of both chemicals in the mixture and the relative percent of one chemical in the mixture to the other. As the antihormone potency of the mixture increases, the slope of the plane increases

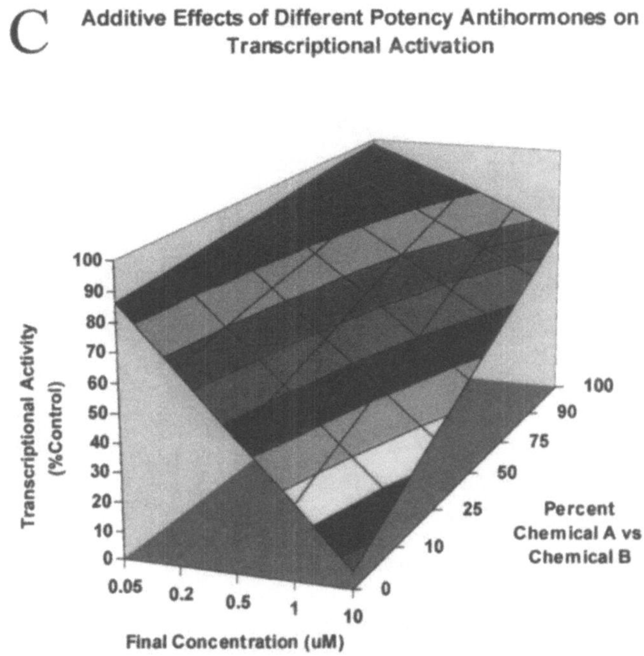

Fig. 2. (continued) C Hypothetical plot of the additive effects of different potency antihormones on transcriptional activation. As the potency of one of the chemicals in the mixture increases relative to the other, the plane slopes toward the more potent chemical. These three-dimensional contour plots in Fig. 2A–C provide a quick method to visually inspect mixture data for additive interactions and relative potency

through the arylhydrocarbon receptor [39], is generally thought to be appropriate for the nuclear steroid hormone receptors, and has recently been applied to environmental endocrine active substances [40, 41], although more than additive [42, 43] and less than additive [44] effects have been observed.

The transcriptional effects of two chemicals interacting at the level of the AR is itself complex. The response can be additive, less than additive, or more than additive compared to either chemical acting alone. The molecular mechanism for the altered response to multiple ligands can involve competition among ligands for monomer binding, the formation of same or mixed-ligand receptor dimers that intrinsically activate or inhibit transcription, and/or the competition of same vs mixed-ligand dimers for binding to regulatory regions of androgen dependent genes, or specific co-activator/co-repressor proteins mandatory for transcriptional activity. The measured effect is thus altered as a result of exposure to multiple chemicals.

To assess the interaction effects of environmental antiandrogen mixtures on transcriptional activity, a Chinese hamster ovary cell line that stably expresses the recombinant human AR together with a mouse mammary tumor virus-driven luciferase reporter was developed. While many different statistical and methodological approaches have been used to analyze chemical interaction

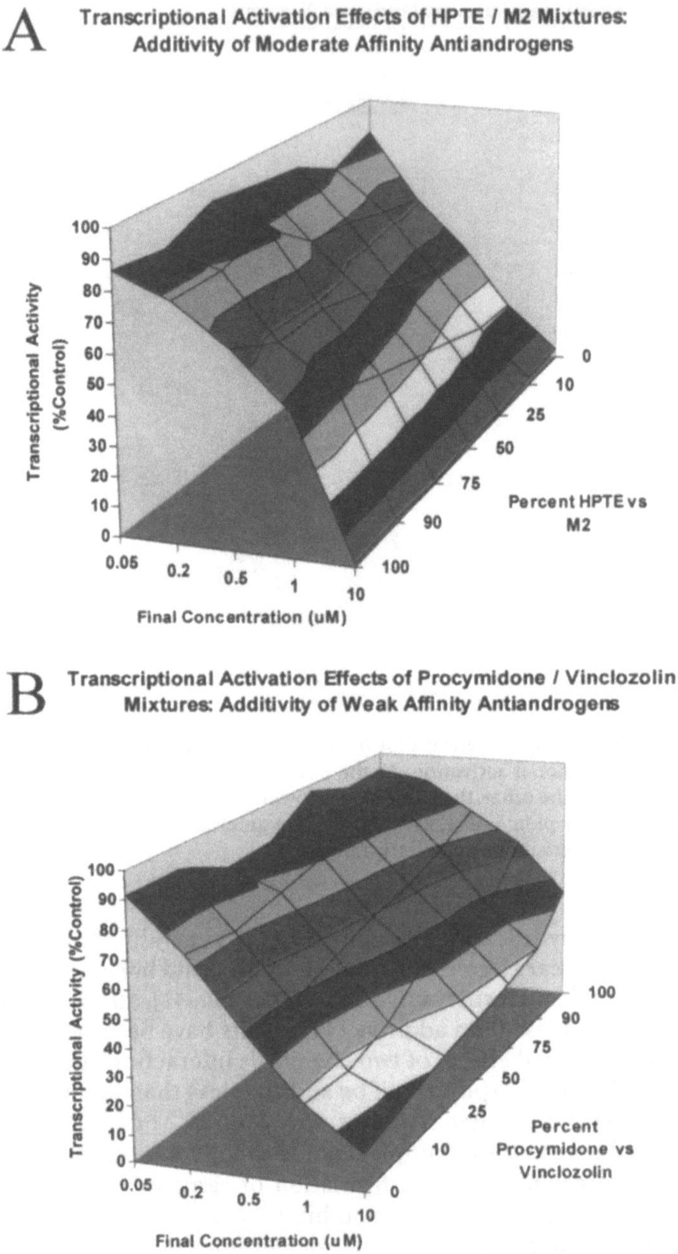

Fig. 3 A, B. Three-dimensional contour plot of empirical data representing the additive effects of: **A** moderately potent; **B** weak antiandrogenic chemicals on transcriptional activation as a function of the final concentration of both chemicals in the mixture and the relative percent of one chemical in the mixture to the other. HPTE and M2 are equipotent antiandrogens with a moderate potency in inhibiting DHT-induced AR-mediated transcriptional activation. The effects of these two chemicals are interpreted to be additive since a plane is produced with a moderately steep slope (**A**). In contrast, vinclozolin and its metabolites are more potent antiandrogens than procymidone or its metabolites, leading to an additive interaction where the plane slopes toward the reader (**B**)

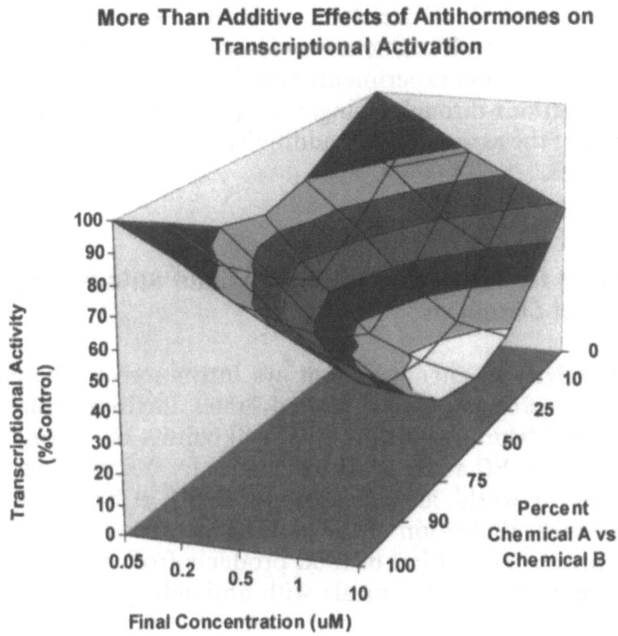

Fig. 4. Three-dimensional contour plot of the effects of two antihormones that interact synergistically to inhibit transcriptional activation. These hypothetical data are plotted as transcriptional activity as a function of the final concentration of both chemicals in the mixture and the relative percent of one chemical in the mixture to the other. A concave surface is produced representing the greater effect on transcriptional activation when both chemicals are present in equimolar amounts. A convex surface would be expected for two chemicals that acted synergistically as receptor agonists

data [45–51], a simple three-dimensional contour plot of the transcriptional activity of two interacting chemicals as a function of both final mixture concentrations and the ratio of one chemical to the other in the mixture is useful for visual inspection of the data. In our studies with environmental antiandrogens, the resulting contour plot approximates a geometric plane where the slope of the plane for two equipotent chemicals reflects total antiandrogenic potency. Flat-sloped planes are indicative of weak antiandrogenic potency (Fig. 2A) and steep-sloped planes indicate strong antiandrogenic potency (Fig. 2B). Mixtures of two chemicals that are not equipotent result in a plane that tilts toward the corner representing the more potent chemical in the mixture (e.g., chemical B in Fig. 2C).

Empirical data from four of the antiandrogenic chemicals discussed above are presented in Fig. 3A, B. Mixtures of moderate affinity equipotent antiandrogens such as HPTE and M2 result in a plane with moderate slope, while interaction effects between the more potent antiandrogen vinclozolin with the less potent procymidone result in a plane tilted toward the reader or toward the corner representing the highest ratio of vinclozolin. The fact that the response surface is a plane suggests that all of the interaction effects of these chemicals on AR-induced transcriptional activity are additive. More than additive antiandrogenic effects would result in a concave response surface similar to that illustrated in Fig. 4. While these experiments have been engineered to be simple in design, i.e., interactions through a single receptor via a common mechanism, the results support the assumption of additivity for simple mixtures of antiandrogenic chemicals.

6 Screening Assays for Measuring AR Agonist and Antagonist Activities of Environmental Chemicals

Detecting ED activity in chemicals that are introduced to the environment has become a pressing issue. The United States Environmental Protection Agency is presently issuing guidelines that will require the testing of new and existing chemicals for ED activity. With evidence for decreasing sperm counts in some parts of the world, an increasing incidence of hypospadias deformity (incomplete masculinization of the external genitalia) in newborn males, and the widespread importation of food products from other countries, it is increasingly important that chemicals with antiandrogen activity be identified.

A number of assays have been developed to determine AR agonist or antagonist activity. These include ligand binding, DNA binding, and transcriptional activation or inhibition assays in yeast or mammalian cells. These approaches provide a first step toward the identification of potential EDs, but must be followed by in vivo whole animal experiments to clarify the role of ligand metabolism and body clearance rates that could activate, inactivate, concentrate, or minimize the parent compound. Environmental antiandrogens identified to date require metabolic conversion to the active forms. It is our experience, however, that some metabolites formed in vivo are recapitulated in cell culture. A potential complication particularly in yeast is that metabolism of the parent compound and the presence of steroid transporters [52] differ between yeast and mammalian cells and tissues in vivo.

6.1 Ligand Binding

Competitive binding assays using [^{3}H]R1881 can establish the apparent equilibrium inhibition constant of unlabeled compounds. Assays are performed using monkey kidney COS cells transfected with the full-length human AR expression vector pCMVhAR [17]. Because the amino acid sequence of the rat and human AR steroid binding domains are identical, similar studies are performed using rat tissue extracts [18]. However, in the latter case, care must be taken to minimize proteolytic breakdown of AR, particularly in androgen responsive tissues such as prostate [53]. Cells are labeled with [^{3}H]R1881 in the presence and absence of test compounds. Inhibition of [^{3}H]R1881 binding using concentrations of the test chemicals up to 50 μmol/l is indicative of AR binding. Higher concentrations can cause cell toxicity and therefore would not reflect binding inhibition.

6.2 DNA Binding and Transcriptional Activation Assays

DNA binding assays make use of baculovirus expressed human AR because sufficient quantities of soluble receptor can be obtained. The use of ^{32}P-labeled, double-stranded oligonucleotides with androgen response element sequence allows for the visualization of AR-DNA complexes in nondenaturing gels [16, 17, 37]. Androgen agonists promote AR DNA binding whereas antagonists disrupt this activity.

Compounds are also tested in transient cotransfection assays in monkey kidney CV1 cells to establish their effect on androgen-induced AR-mediated transcriptional activity. Alternatively, human cells that express the androgen receptor endogenously can be transfected with the reporter to assess effects on transcriptional activity. By including the test compound in the presence of 0.1 nmol/l DHT, antagonist activity is assessed. Controls for cellular toxicity include the use of the constitutively active AR NH_2-terminal and DNA binding domain fragment AR1–660, and the glucocorticoid receptor [18], or a constitutively active reporter. No change in these activities indicates a compound is probably not toxic to cells and transcriptional inhibition results from antagonist activity. Agonist activity is tested in the same cotransfection assay except incubations are performed in the absence of DHT. We typically observe at least 40-fold induction of luciferase activity with 0.1 nmol/l DHT using cell lines that express the receptor endogenously or in CV-1 cells using 25 ng of the AR expression vector and 5 μg of the mouse mammary tumor virus promoter linked to the firefly luciferase reporter gene using a calcium phosphate-DNA precipitation transfection method. Metabolism studies are required to establish the stability of the compounds during in vitro cell culture.

6.3 N/C Interaction Assay

The NH_2- and carboxyl-terminal regions of the AR interact [13, 14] in such a way as to augment the agonist activity of low physiological androgen concentrations [15]. This conclusion is based on the observation that deletion of the NH_2-terminal region causes an increase in the dissociation rate of bound androgen without altering the equilibrium binding affinity [12], suggesting that the association rate is similarly increased. A direct test of the NH_2-/carboxyl-terminal (N/C) interaction using a two-hybrid protein interaction assay in Chinese hamster ovary cells revealed that it depends on androgen binding to the ligand binding domain [13]. Specific mutations in the AR gene that cause androgen insensitivity have provided evidence that the N/C interaction occurs intermolecularly, is associated with receptor dimerization [14], and has its interaction site in the activation function 2 region of the AR ligand binding domain [54]. The N/C interaction is inhibited by lower affinity antagonists such as hydroxyflutamide and those found to be active in the environment due to pesticide or fungicide use. However, high affinity antagonists with partial agonist activity may disrupt the N/C interaction. In this case, agonist activity occurring in the absence of the N/C interaction would require higher hormone concentrations through an as yet unknown mechanism. The influence of AR antagonists on this intermolecular interaction may be indicative of their activity in vivo.

Previous androgen agonist and antagonist assays made use of the entire AR coding region of the AR gene. In contrast, the N/C assay is based on the molecular interaction between two domains of the AR that is specific for androgen binding and disrupted by antagonists. The N/C interaction can be performed by cotransfection of mammalian or yeast cells. The human AR ligand binding domain residues 624–919 are expressed as a fusion protein with the GAL4 DNA binding domain, and the AR NH_2-terminal and DNA binding domain residues 1–660 are expressed as a fusion protein with the VP16 transactivation domain. The reporter vector G5E1bLuc contains the firefly luciferase coding sequence preceded by 5 GAL4 binding sites and the E1bTATA promoter. Because there are numerous interactions involved in transcriptional activation by the full-length AR, this assay was developed to focus on a single androgen specific interaction that is sensitive to endocrine disruption. The assay can be scaled up to screen large numbers of samples using automation.

To establish whether this mammalian cell-based assay is effective as a screening method for environmental chemicals with androgen agonist or antagonist activity, we tested the compounds listed below with known or unknown AR activities to establish the in vitro criteria necessary to predict in vivo agonist and antagonist activity. The N/C interaction is induced by 0.1–1 nmol/l DHT, testosterone, or by the synthetic androgens methyltrienolone and mibolerone. Also, the much lower affinity anabolic steroids oxandrolone and fluoxymesterone induce the N/C interaction [15].

Shown in Table 2 are the approximate concentrations required for 50% inhibition of the N/C interaction induced by 1 nmol/l DHT. The dose dependence of inhibition of the N/C interaction correlates with AR antagonist activity, with the most potent antagonists, such as hydroxyflutamide, inhibiting the DHT-induced N/C interaction at lower concentrations. The concentration dependence of inhibition of the N/C interaction appears to provide a measure for in vivo AR antagonist activity. The most effective environmental inhibitors, HPTE, M2, and *p,p′*-DDE, were shown previously to be environmental antiandrogens [17, 18, 37]. The AR N/C screen will therefore be useful in testing additional compounds for AR antagonist activity. For the environmental compounds, inhibition of the N/C interaction paralleled the antagonist potency determined previously in vivo.

Recent studies indicate, however, that compounds that bind the AR with high affinity and have agonist activity in transient transfection assays do not necessarily display agonist activity in vivo. This was demonstrated most strikingly with medroxyprogesterone acetate (MPA), available for clinical use as Provera. MPA is a weak androgen in vivo that binds AR with high affinity but is the most effective inhibitor of the N/C interaction yet identified. The N/C interaction in-

Table 2. Concentration dependence of inhibition of the AR NH_2- and carboxyl-terminal (N/C) interaction. The N/C interaction was performed in CHO cells by transient transfection of plasmids expressing the AR NH_2-terminal 1–660 amino acid residues as a fusion protein with the VP16 transactivation domain, and the AR carboxyl-terminal ligand binding domain 624–919 amino acid residues as a fusion protein with the GAL4 DNA binding domain as previously described [13, 14]. Shown are the approximate concentrations of the indicated compounds that reduced the N/C interaction by 50% in the presence of 1 nmol/l DHT. ND denotes no detectable inhibition when tested up to concentrations of 10 µmol/l test compound

Ligand concentrations that inhibit the N/C interaction induced by 1 nmol/l DHT			
Hydroxyflutamide	50 nmol/l	Bisphenol A	10 µmol/l
Cyproterone acetate	50 nmol/l	Dibutyl phthalate	10 µmol/l
HPTE	200 nmol/l	*o,p'*-DDT	ND
M2	500 nmol/l	Benzyl butyl phthalate	ND
p,p'-DDE	500 nmol/l	Iprodione	ND
Vinclozolin	1 µmol/l	Kelthane	ND
Methoxychlor	1 µmol/l	Bisphenol A 3EO	ND
Octylphenol	1 µmol/l	Bisphenol A 2EO	ND
Procymidone	5 µmol/l	Butylphenol	ND
M1	10 µmol/l	Nonylphenol	ND

duced by 1 nmol/l DHT is inhibited 50% by 1 nmol/l MPA whereas 50 times higher concentrations are required for similar inhibition by hydroxyflutamide. MPA activates luciferase activity in transient assays at concentrations of 0.1 nmol/l, about 100-fold higher concentration than for similar activation by DHT. Failure of MPA to induce the N/C interaction may result in the higher concentrations necessary to activate AR. This was in agreement with the higher concentrations of MPA required for AR stabilization against degradation. It indicates, however, that sufficiently high levels of a ligand can induce AR agonist activity even if it does not induce the N/C interaction. The results suggest that the N/C interaction facilitates agonist activity at the low circulating androgen concentrations during fetal development and later in life.

Until recently, no environmental AR agonists have been identified. None of the compounds listed above induce the N/C interaction in the absence of androgen, suggesting they lack AR agonist activity in vivo, in agreement with in vivo studies. However, such compounds likely exist in the environment because of the report of a female to male shift in sex ratios among mosquitofish observed downstream of paper mills [55]. While preliminary evidence suggested a possible role of β-sitosterol, a steroid produced by bacterial degradation of wood products and identified in the effluent of paper mills, it appears more likely that steroidal metabolites of β-sitosterol account for the androgen activity. Studies are in progress that make use of the techniques described above to identify what would be the first environmental androgen that would account for the masculinization of exposed female fish.

6.4 In Vivo Assays

In vivo assays recommended by EDSTAC to screen environmental chemicals for antiandrogen activity include the 7-day Hershberger assay, a 15-day adult male, and a 28-day pubertal male repeated dosing assay. However, the 15-day adult male assay fails to detect the antiandrogen effects of *p,p'*-DDE in some rat strains [56] and the Hershberger assay does not detect the antiandrogen activity of dibutyl phthalate [57]. In contrast, the pubertal male rat assay detects antiandrogen activity for both chemicals [58] suggesting it is more predictive than the other in vivo assays.

Balano-preputial separation in the male pubertal assay for antiandrogen activity is also the endpoint for the Office of Economic Cooperation and Development Test Guideline Protocol (proposed revisions to OECD 416) for the High-Production Volume Program and the EPA Health Effects Test Guidelines (870.3800) for the Children's Health Test Rule Program. Assessment of this same endpoint using the 28-day pubertal male assay in the EPA ED screening program would help to standardize these assessment protocols. Similarly, the 28-day pubertal female assay reveals the age at vaginal opening as the endpoint for detecting chemicals with estrogen activity in the High-Production Volume and Children's Health Test Rule Programs. With the detection of antithyroid hormone activity included in both pubertal tier I screening assays, selection of these two pubertal screening assays for the ED screening program would move toward standardizing these three screening programs for the three hormone activities.

Potential drawbacks of these in vivo assays are that preputial separation may not be as sensitive as assessing sex accessory organ weight in the Hershberger assay. However, preputial separation as an endpoint will likely drive risk assessment of environmental antiandrogens via multigenerational assays, so its use in screening assays is appropriate. Opposition to this recommendation will likely be based on objections to using screening data for risk assessment, which was not the intent of the EDSTAC recommended tier I assays. To avoid use of inappropriate screening data for risk assessment, additional dose groups could be used where animals are dosed by the most appropriate route and over a wide dose response range in addition to the route and dose range recommended by EPA for the tier I screening assay. In this way, issues related to the relevance of the route of chemical administration and dose level could be included in the risk assessment process. For pesticide chemicals supported by past multigenerational studies under outdated EPA guidelines, it would not be unreasonable to use the results from these antiandrogen screening assays in lieu of a full multigenerational study, since preputial separation is the new endpoint assessed in the new guidelines that detects antiandrogen activity. The burdens placed on industry and government for ED screening and testing are challenging and require the conservation of efforts and standardization of protocols where possible.

7 Conclusions

With the aim toward completing the EDSTAC recommendations to broaden the scope of potential ED activity and the affected species, a series of potential assay protocols for detecting estrogen, androgen, and thyroid hormone activities is being outlined by the EPA EDSP Task Force. Once a consensus is reached, protocols will be standardized and validated through a multi-chemical and multi-laboratory validation process. Standardized and validated assays will be used to assess ED activity of a number of marketed chemicals that will be prioritized based on production volume, potential for human exposure, existing ED data, and results of the high-throughput prescreening effort. Standardizing endocrine testing strategy, criteria, and procedures will make possible the identification of potentially hazardous chemicals with ED activity. Furthermore, continued basic research, like that described here for the AR, will be pivotal in the development of simpler and more comprehensive screening assays, together with the identification of additional receptor and/or developmental systems for which screening is necessary to protect human and ecological health.

Acknowledgments. We are grateful to Christy Lambright and Kathy Bobseine, Reproductive Toxicology Division, National Health and Environmental Effects Research Laboratory, United States Environmental Protection Agency, Research Triangle Park, North Carolina, for completing the additivity studies with M2, HPTE, vinclozolin, and procymidone, and to Jon A. Kemppainen, University of North Carolina, Chapel Hill, North Carolina, for the analysis of β-sitosterol activity. This work was supported by Grant ES-08265 from the National Institute of Environmental Health Sciences of the National Institutes of Heath, USA.

8 References

1. Guillette LJ, Gross TS, Masson GR, Matter JM, Percival HF, Woodward AR (1994) Environ Health Perspect 102:680
2. Schardein J (1993) In: Chemically induced birth defects, 2nd edn. Dekker, New York, chap 9, p 271
3. Kelce WR, Gray LE (1997) Health Environ Digest 11:9
4. Peterson R, Cooke P, Gray LE, Kelce WR (1997) In: Boekelheide K, Chapin R (eds) Reproductive and endocrine toxicology: male reproductive toxicology, vol 10. Elsevier Science, New York, p 181
5. Gray LE, Monosson E, Kelce WR (1996) In: Di Giulio R, Monosson E (eds) Interconnections between human and ecosystem health. Chapman and Hall, New York, p 46
6. Guillette LJ, Crain DA (1996) Com Toxicol 5:38
7. Guillette LJ, Arnold SF, McLachlan JA (1996) Animal Reprod Sci 42:13
8. Toppari J, Larsen JC, Christiansen P, Giwercman A, Grandjean P, Guillette LJ (1995) Miljoprojekt nr 290:1
9. Kelce WR, Wilson EM (1997) J Mol Med 75:198
10. Kavlock RJ, Daston GP, DeRosa C, Fenner-Crisp P (1996) Env Health Persp 104:715
11. Wilson EM, French FS (1976) J Biol Chem 251:5620
12. Zhou ZX, Lane MV, Kemppainen JA, French FS, Wilson EM (1995) Mol Endocrinol 9:208
13. Langley E, Zhou ZX, Wilson EM (1995) J Biol Chem 270:29,983

14. Langley E, Kemppainen JA, Wilson EM (1997) J Biol Chem 273:92
15. Kemppainen JA, Langley E, Wong CI, Bobseine K, Kelce WR, Wilson EM (1999) Mol Endocrinol 13:440
16. Wong CI, Zhou ZX, Sar M, Wilson EM (1993) J Biol Chem 268:19,004
17. Wong CI, Kelce WR, Sar M, Wilson EM (1995) J Biol Chem 270:19,998
18. Kelce WR, Stone CR, Laws SC, Gray LE, Kemppainen JA, Wilson EM (1995) Nature 375: 581
19. Sinclair AH, Berta P, Palmer MS, Hawkins JR, Griffiths BL, Smith MJ, Foster JW, Frischauf AM, Lovell-Badge R, Goodfellow PN (1990) Nature 346:240
20. Wilson JD (1978) Ann Rev Physiol 40:279
21. Lubahn DB, Joseph DR, Sullivan PM, Willard HF, French FS, Wilson EM (1988) Science 240:327
22. Quarmby VE, Kemppainen JA, Sar M, Lubahn DB, French FS, Wilson EM (1990) Mol Endocrinol 4:1399
23. Edwards DP, Altmann M, DeMarzo A, Zhang Y, Weigel NL, Beck CA (1995) J Steroid Biochem Mol Biol 53:449
24. Melamed M, Castano E, Notides AC, Sasson S (1997) Mol Endocrinol 11:1868
25. Gray LE, Ostby JM, Marshall R (1993) Biol Reprod 48[Suppl 1]:97 abst 154
26. Gray LE, Ostby JS, Kelce WR (1994) Toxicol Appl Pharmacol 129:46
27. Kelce WR, Monosson E, Gamcsik MP, Laws SC, Gray LE (1994) Toxicol Appl Pharmacol 126:276
28. Hosokawa S, Murakami M, Ineyama M, Yamada T, Yoshitake A, Yamada H, Miyamoto J (1993) J Toxicol Sci 18:83
29. Hosokawa S, Murakami M, Ineyama M, Yamada T, Koyama Y, Okuno Y, Yoshitake A, Yamada H, Miyamoto J (1993) J Toxicol Sci 18:111
30. Ostby J, Kelce WR, Lambright C, Wolf CJ, Mann P, Gray LE (1999) Tox Ind Health 15:80
31. Imperato-McGinley J, Sanchez RS, Spencer JR, Yee B, Vaugan ED (1992) Endocrinology 131:1149
32. Nelson JA, Struck RF, James R (1978) J Toxicol Environ Health 4:325
33. Bulger WH, Muccitelli RM, Kupfer D (1978) J Toxicol Env Health 4:881
34. Kupfer D, Bulgerm WH (1987) Fed Proc 46:1864
35. Gray LE, Ostby J, Ferrell J, Rhenberg G, Linder R, Cooper R, Goldman J, Slott V, Laskey J (1989) Appl Toxicol 12:92
36. Eroschenko VP, Atta AA, Grober MS (1995) Reprod Toxicol 9:379
37. Kelce WR, Lambright CR, Wiese TE, Gray LE, Wong CI, Wilson EM (submitted)
38. Safe S (1990) CRC Crit Rev Toxicol 21:51
39. Zabel EW, Walker MK, Hornung MW, Clayton MK, Peterson RE (1995) Toxicol Appl Pharmacol 134:204
40. Safe SH (1998) Environ Health Perspec 106[Suppl 4]:1051
41. Gao X, Son D-S, Terranova PF, Rozman KK (1999) Toxicol Appl Pharmacol 157:107
42. Arnold SF, Vonier PM, Collins BM, Klotz DM, Guillette LJ (1997) Environ Health Perspect 105[Suppl 3]:615
43. Arnold SF, Bergeron JM, Tran DQ, Collins BM, Vonier PM, Crews D, Toscano WA, McLachlan JA (1997) Biochem Biophy Res Comm 235:336
44. Safe SH (1994) Environ Health Perspec 103:346
45. Yang RSH (1998) Environ Health Perspec 106[Suppl 4]:1059
46. Chou TC, Talalay P (1981) Eur J Biochem 115:207
47. Chou TC, Talalay P (1984) Adv Enzyme Regul 22:27
48. El-Masri HA, Reardon KF, Yang RSH (1997) CRC Crit Rev Toxicol 27:175
49. Carter WH, Wampler GL (1986) Cancer Treat Rep 70:133
50. Carter WH, Ginnings C, Staniswalis JG, Campbell ED, White KL (1988) J Am College Toxicol 7:963
51. Carter WH, Ginnings C (1994) In: Yang RSH (ed) Toxicology of chemical mixtures: case studies, mechanisms, and novel approaches. Academic Press, San Diego, p 643
52. Kralli A, Yamamoto KR (1996) J Biol Chem 271:17,152

53. Wilson EM, French FS (1979) J Biol Chem 254:6310
54. He B, Kemppainen JA, Voegel JJ, Gronemeyer H, Wilson EM (1999) J Biol Chem 274:37,219
55. Denton TE, Howell WM, Allison JJ, McCollum J, Marks B (1985) Bull Environ Contam Toxicol 35:627
56. O'Connor JC (1999) DuPont Haskell Lab, personal communication
57. Gray LE (1999) US EPA, personal communication
58. Gray LE, Wolf C, Lambright C, Mann P, Price M, Cooper RL, Ostby J (1999) Toxicol Ind Health 15:94

CHAPTER 4

Chemistry of Natural and Anthropogenic Endocrine Active Compounds

Manfred Metzler*, Erika Pfeiffer

Institute of Food Chemistry and Toxicology, University of Karlsruhe, P.O. Box 6980, 76128 Karlsruhe, Germany
e-mail: manfred.metzler@chemie.uni-karlsruhe.de

Research over the past years has revealed that numerous compounds present in our environment exert hormonal activity and thus have the potential to interfere with the endocrine system of humans and animals. These endocrine active compounds comprise both naturally occurring substances and man-made chemicals, and their chemical structures are surprisingly diverse. This chapter provides an overview of the chemical structures of typical endocrine active compounds, which are discussed in other chapters of this book with respect to their occurrence, mechanisms of action, implications for human health, and environmental significance.

Keywords. Endocrine active compounds, Chemical structures, Phytoestrogens, Industrial chemicals

1	**Introduction**	64
2	**Sex Hormones and Natural Sex Hormone-Like Compounds**	65
2.1	Estrogens	65
2.1.1	Mammalian Estrogens	65
2.1.2	Phytoestrogens	66
2.1.3	Mycoestrogens	69
2.2	Androgens and Progestins	70
3	**Anthropogenic Sex Hormone-Like Compounds**	71
3.1	Pharmaceutical Compounds	71
3.1.1	Estrogens and Antiestrogens	71
3.1.2	Androgens and Antiandrogens	73
3.1.3	Progestins and Antiprogestins	73
3.2	Industrial Chemicals	73
3.2.1	Phenolic Compounds	74
3.2.2	Polychlorinated Compounds	76
3.2.3	Phthalates	76
3.2.4	Miscellaneous Compounds	78
4	**Conclusion**	79
5	**References**	80

The Handbook of Environmental Chemistry Vol. 3, Part L
Endocrine Disruptors, Part I
(ed. by M. Metzler)

Abbreviations

BADGE	bisphenol A diglycidyl ether
BBP	butyl benzyl phthalate
BPA	bisphenol A
BPA-DMA	bisphenol A dimethacrylate
DAS	4,4′-diaminostilbene-2,2′-disulfonic acid, amsonic acid
DBCP	1,2-dibromo-3-chloropropane
p,p′-DDE	2,2-bis(*p*-chlorophenyl)-1,1-dichloroethene
o,p′-DDT	2-(*o*-chlorophenyl)-2-(*p*-chlorophenyl)-1,1,1-trichloroethane
p,p′-DDT	2,2-bis(*p*-chlorophenyl)-1,1,1-trichloroethane
DEHP	di(2-ethylhexyl) phthalate
DES	diethylstilbestrol
DHT	5α-dihydrotestosterone
E1	estrone
E2	17β-estradiol
E3	estriol
EAC	endocrine active compound
HPTE	2,2-bis(*p*-hydroxyphenyl)-1,1,1-trichloroethane
NDGA	nordihydroguaiaretic acid
PCB	polychlorinated biphenyls
PCDD	polychlorinated dibenzo(*p*)dioxins
PCDF	polychlorinated dibenzofurans

1 Introduction

Numerous physiological processes are regulated by endocrine hormones, which are delivered from their tissues of production to their target organs via the blood stream [1]. One important group within the large and diverse family of endocrine hormones are sex hormones, which play multiple and important roles in mammalian organisms during virtually every stage of life from conception to old age, as will be discussed in several chapters of this book. Over the past years, evidence is accumulating that numerous compounds present in our environment have the potential to interfere with the actions of endogenous hormones [2]. These compounds are called endocrine active compounds (EAC). Most EAC known to date affect the target tissues of sex hormones, although other organs may also be affected. It is usually believed that EAC cause adverse effects. This has expanded the interest in sex hormones and their corresponding EAC from medicine and basic research into the public and political arena.

The large number of sex hormone-related EAC and their structural diversity are confusing to the layman and sometimes even to scientists. The purpose of this chapter is to review the most pertinent chemical structures. Many EAC with sex hormone activity are naturally occurring, whereas others are man-made. The number of individual compounds in both categories will undoubtedly grow over the coming years, as assays for detecting EAC are improved and widely used.

2
Sex Hormones and Natural Sex Hormone-Like Compounds

In this chapter, the term sex hormones will be reserved for the physiological compounds formed in endocrine glands, sometimes also called endogenous sex hormones. Other naturally occurring substances able to mimic or antagonize the action of the endogenous sex hormones are termed sex hormone-like compounds. According to their physiological effects, sex hormones and sex hormone-like compounds are classified as estrogens, androgens or progestins.

2.1
Estrogens

2.1.1
Mammalian Estrogens

In mammalian organisms including humans, the physiological estrogens are steroids derived from 5α-estrane. 17β-Estradiol (E2, Fig. 1) is the steroidal estrogen with the highest activity. Characteristic chemical features of E2 are the aromatic ring (ring A) with a hydroxy group at C3, and a second hydroxy group at the C17 position of ring D. The formula of E2 in Fig. 1 also indicates the conformation of rings B, C, and D, and the orientation of the C17 hydroxy group. It is common to use simplified formulas not showing the stereochemistry of the ring system, such as for the two other important steroidal estrogens estrone (E1) and estriol (E3). E3 and 16α-hydroxy-E1 are major metabolites of E2 and E1, respectively, but hydroxylations at other positions also take place in the metabolism of E2 and E1, e.g., at C2, C4, C6, etc.

In the urine of pregnant mares, the steroidal estrogens equilin and equilenin (Fig. 1) are excreted in significant amounts. These estrogens are widely used as

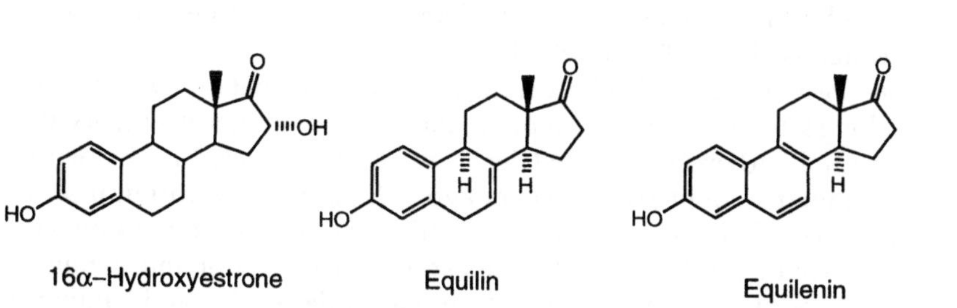

Fig. 1. Major mammalian steroidal estrogens

estrogen replacement for the treatment of menopausal symptoms, e.g., as the drug Premarin.

2.1.2
Phytoestrogens

It has long been known that certain plants contain compounds able to mimic the biological effects of steroidal estrogens [3]. Today, such phytoestrogens have been detected in more than 300 plants, some of which (like soy beans) are widely used for human nutrition [4]. Until recently, the chemical structures of about twenty phytoestrogens were elucidated and categorized into the three classes isoflavones, coumestans and lignans. Due to the present interest in environmental EAC, new phytoestrogens either within these classes or with novel structures are being identified at an accelerated rate. Some examples will be given below.

Although the hormonal activity of the phytoestrogens known to date is two to five orders of magnitude below that of E2, their high concentration in certain plants and their slower metabolic disposition can lead to tissue levels exceeding those of the endogenous estrogens by a factor of thousand or more (see Chap. Dietary Estrogens of Plant and Fungal Origin: Occurrence and Exposure (Part I)). At these concentrations, even the low intrinsic activity of phytoestrogens may suffice to cause hormonal effects.

Isoflavones, which share the 3-phenylchromone ring system, represent the largest group of phytochemicals with estrogenic activity. Important members of this group are displayed in Fig. 2. Their occurrence, effects and association with disease are dealt with in Chap. Dietary Estrogens of Plant and Fungal Origin: Occurrence and Exposure (Part I) and Beneficial and Adverse Effects of Dietary Estrogens on the Human Endocrine System: Clinical and Epidemiological Date (Part II). Daidzein, genistein and, to a lesser extent, glycitein are common in soy and many soy-derived food items. In general, isoflavone phytoestrogens are present as glycosides in the plant and are hydrolyzed to the free aglycones either during processing of the food (e.g., in fermented soy products) or after ingestion in the intestine. Intestinal bacteria are thought to play a major role in glycoside hydrolysis. Another important metabolic reaction of the gut flora is the reduction of the isoflavone ring, converting for example daidzein to equol and *O*-desmethylangolensin.

The major representative of the coumestane phytoestrogens, which all contain a 2-phenylcoumarin ring system, is coumestrol (Fig. 3). Coumestrol exhibits the highest hormonal activity of all known phytoestrogens in most assay systems, which is about 1% that of E2.

A more recent class of phytoestrogens is represented by the lignans. Although long known as important phytochemicals of numerous plants, the lignans enterodiol and enterolactone (Fig. 4) were first detected in the urine and plasma of humans about twenty years ago [5]. Shortly thereafter, it was disclosed that enterodiol and enterolactone, which are now called mammalian lignans, are derived from the plant lignans secoisolariciresinol and matairesinol [6]. The latter occur in numerous plants, in particular grains and legumes (see Chap. Dietary Estrogens of Plant and Fungal Origin: Occurrence and Exposure

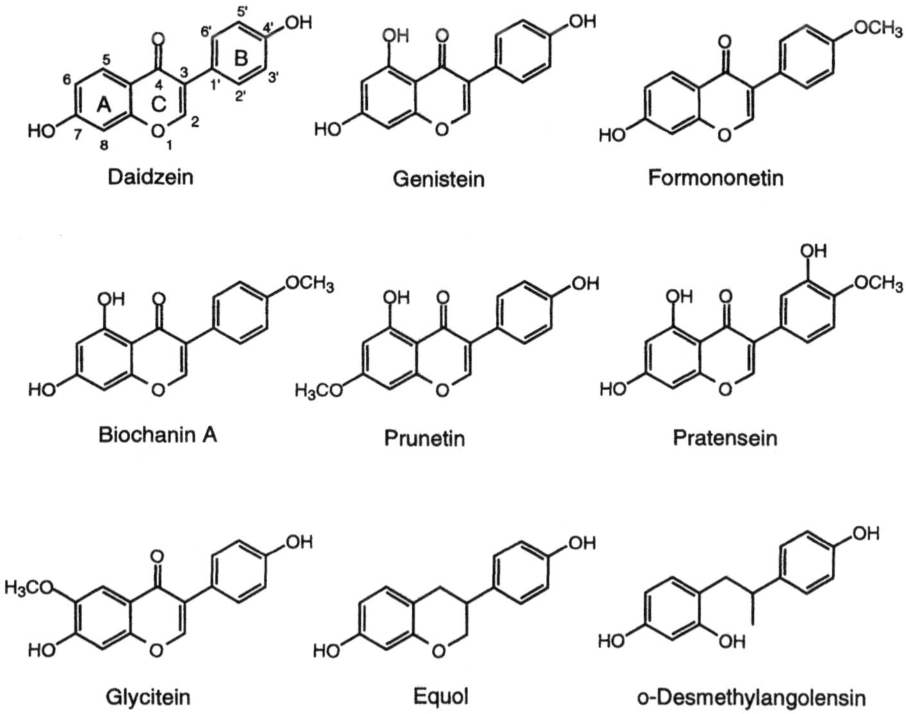

Fig. 2. Chemical structures of common isoflavone phytoestrogens

(Part I)). The highest concentrations of secoisolariciresinol and matairesinol are found in flaxseed. Other lignans recently identified in flaxseed are isolariciresinol and pinoresinol, whereas lariciresinol and nordihydroguaiaretic acid (NDGA) were not detected [7]. NDGA and guaiaretic acid are major constituents of chaparral, a desert shrub dwelling in the southwestern USA and Mexico, and used in folk medicine for herbal infusions [8].

Lignans have two or more chiral carbon atoms and may exist as enantiomers or diastereomers. The stereochemistry has been well studied for plant lignans in conjunction with their biosynthesis [9] but not for mammalian lignans. Therefore, the chemical formulas of mammalian lignans (Fig. 4) are based on the assumption that the chirality of the plant lignans is not changed during their bacterial conversion to mammalian lignans.

Recently, it has been disclosed that some flavonoids have estrogenic activity [10]. Flavonoids comprise flavones (containing the 2-phenylchromone moiety) like apigenin, flavonols (containing the 2-phenyl-3-hydroxychromone moiety) like kaempferol, flavanones (containing the 2-phenylchromanone moiety) like naringenin, and flavanols (containing the 2-phenyl-3-hydroxychromane moiety) like catechin and epicatechin (Fig. 5). Flavonoids represent a very large group of phytochemicals and are found in numerous plants. 8-Prenylnaringenin (Fig. 5) is one of the more recently identified phytoestrogens, shown to occur in high concentrations in hops and in low concentrations in beer [11]. Surprisingly, 8-prenyl-

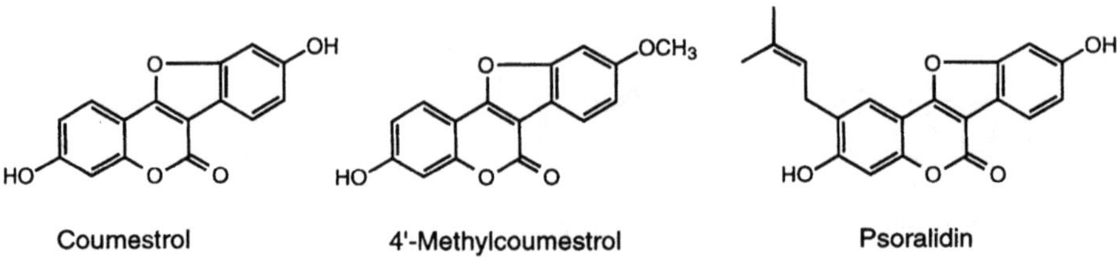

Fig. 3. Chemical structures of coumestane phytoestrogens

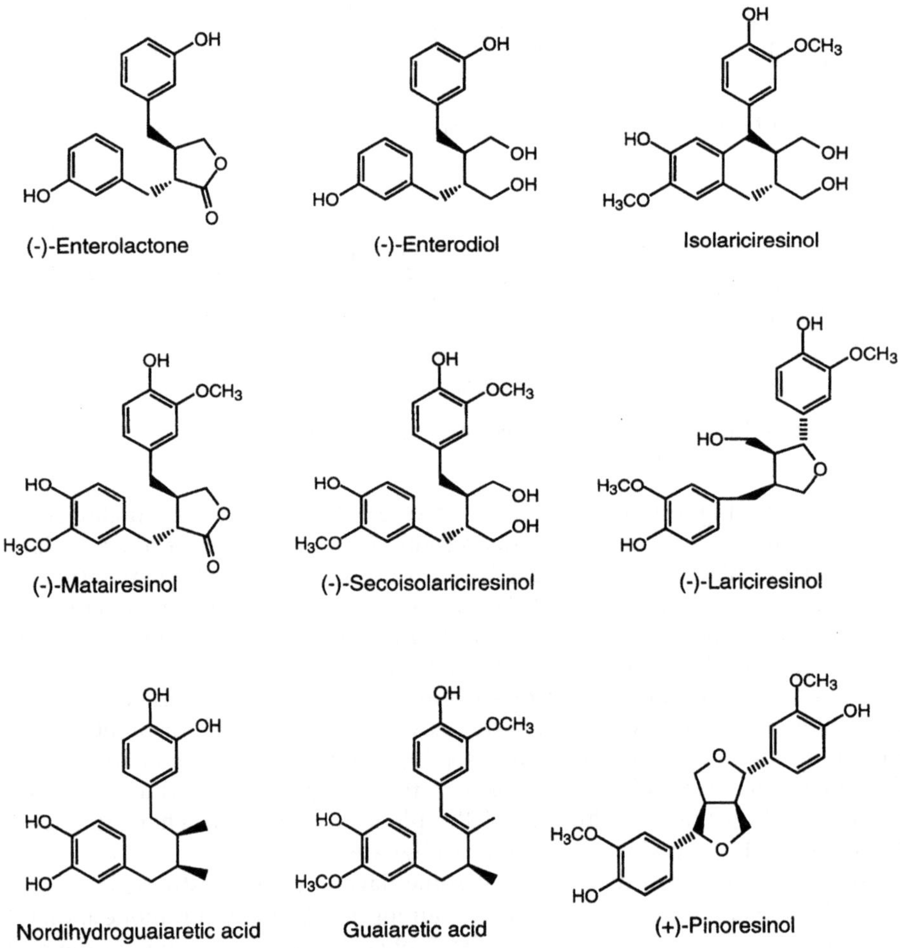

Fig. 4. Chemical structures of lignan phytoestrogens

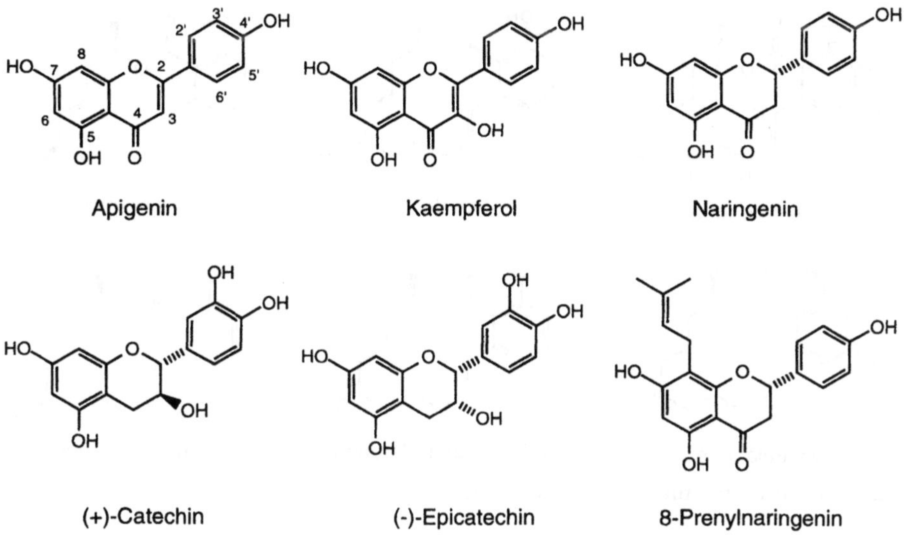

Fig. 5. Chemical structures of flavonoids

naringenin proved to be more estrogenic than any other known phytoestrogen [11]. Catechin is a typical flavonoid of red wine [12], which has recently been associated with reduced rates of coronary heart disease mortality [13].

Finally, hydroxylated stilbenes, another well-known class of phytochemicals, have recently gained renewed interest because of the putative beneficial health effects of some of their congeners [14]. Typical compounds are resveratrol and piceatannol (Fig. 6), the *trans*- and *cis*-isomers of which occur in red wine as glucosides at the 3-hydroxy group [15]. The glucoside of resveratrol is named piceid and that of piceatannol is called astringin. Lower concentrations of piceid and astringin were also detected in white wines [15].

2.1.3
Mycoestrogens

Several strains of *Fusarium* molds, which frequently infest corn (*Zea mays*) and other crops, produce a resorcylic acid lactone called zearalenone (Fig. 7). Zearalenone is held responsible for the fertility problems and estrogenization of

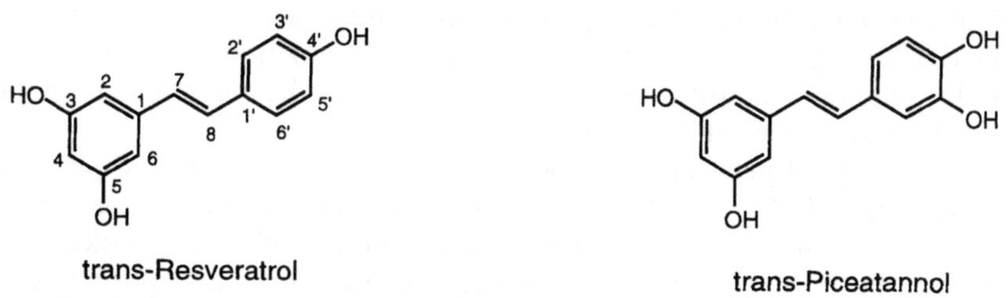

Fig. 6. Chemical structures of stilbene phytochemicals

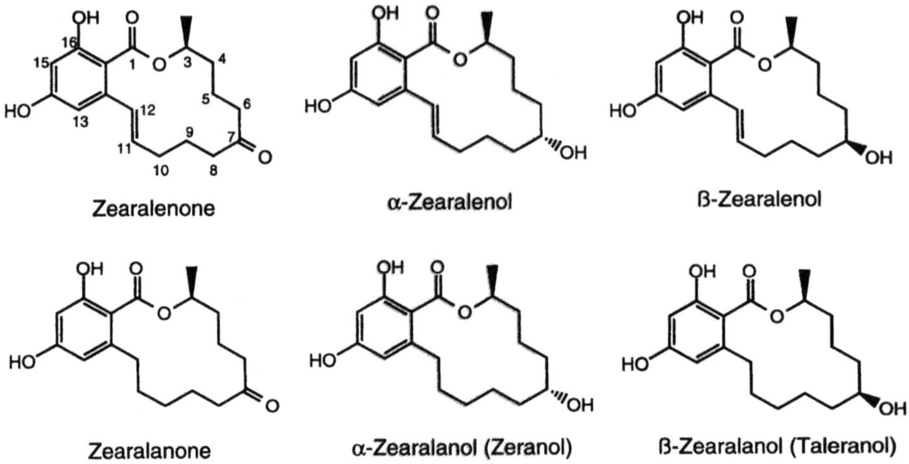

Fig. 7. Chemical structures of mycoestrogens

farm animals, in particular pigs, observed after feeding of contaminated hay or corn ("moldy corn disease", see Chap. Dietary Estrogens of Plant and Fungal Origin: Occurrence and Exposure (Part I)). The name zearalenone indicates both the biological source of this estrogenic fungal product as well as some of its characteristic structural features, i.e., the double bond at C11 and the keto group at C7. Reduction of the keto group gives rise to *α*- and *β*-zearalenol, whereas reduction of the double bond leads to zearalanone (Fig. 7). The products of reduction of both groups are *α*- and *β*-zearalanol. *α*-Zearalanol (zeranol) is more estrogenic than its *β*-isomer (taleranol), and is used as a growth promotor in cattle in the USA and other countries (trade names Ralgo®), Ralabol® and others). Zeranol is also formed in cattle, probably in the rumen, from zearalenone and even to a higher extent from *α*-zearalenol but not from *β*-zearalenol after ingestion [16]. *α*-Zearalenol and *β*-zearalenol are products of the fungal synthesis of resorcylic acid lactones as well as bovine metabolites of zearalenone.

2.2 Androgens and Progestins

The natural sex hormone of males, produced mainly in the Leydig cells of the testis, is testosterone, a steroid with 19 carbon atoms (Fig. 8). In most androgen target tissues, e.g., the prostate, the hormonally more active 5*α*-dihydrotestosterone (DHT) is generated from testosterone by the enzyme 5*α*-reductase. See textbooks on endocrinology, e.g., [1], for details about the biosynthesis, metabolism and physiological effects of androgens.

The physiological progestin is progesterone (Fig. 8), a C_{21}-steroid produced in females mainly in the follicle after ovulation. Most biological effects of progesterone depend on the presence of estrogens. As outlined in textbooks on endocrinology, both the ratio of progesterone to estradiol concentration and the sequence of action of the two hormones are of importance for the effects.

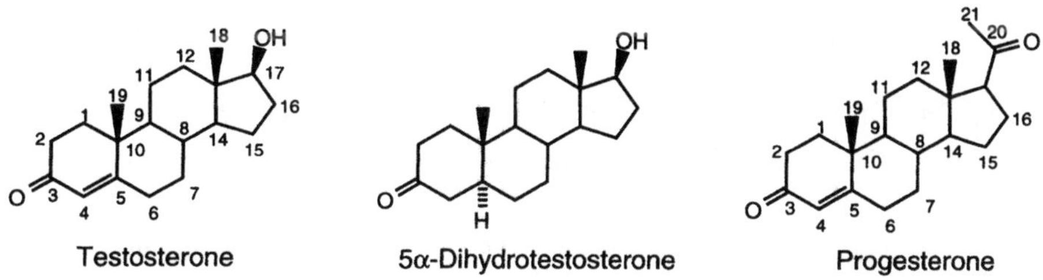

Fig. 8. Chemical structures of the mammalian androgens and progestins

3 Anthropogenic Sex Hormone-Like Compounds

Man-made compounds mimicking or antagonizing the effects of the endogenous sex hormones have been designed and produced by pharmaceutical companies for several decades for a variety of uses in human and veterinary medicine. Examples for such applications of synthetic hormone-like compounds are the substitution of endogenous hormones of women after menopause or in certain diseases, the synchronization of estrus in farm animals, or the prevention or stimulation of conception. Anthropogenic sex hormone-like compounds of this "pharmaceutical" type are generally characterized by a high intrinsic activity and oral efficacy. Only recently has this group been joined by a growing number of anthropogenic compounds produced for completely different purposes, e.g., as pesticides or plastic monomers, but fortuitously also possessing hormonal activity. These "industrial" sex hormone-like compounds exhibit, in most cases, very low intrinsic hormonal activity but are often persistent in the body and environment. Although this latter group enjoys a lot of scientific and public attention, it should not be forgotten that the "pharmaceutical" hormone-like compounds are still produced and applied on a large scale, and that the waste of their production and consumption also enters the environment.

3.1 Pharmaceutical Compounds

Most of the pharmaceutical-type of sex hormone-like compounds are derived from the natural endogenous hormones and still contain the steroid moiety. Only in the case of man-made estrogenic substances, have other chemical classes been successfully synthesized, e.g., stilbene-type estrogens and triphenylethylene-type antiestrogens.

3.1.1 *Estrogens and Antiestrogens*

Representative examples of steroidal and stilbene-type estrogens and triphenylethylene-type antiestrogens are depicted in Fig. 9. The steroid 17α-ethinylestradiol has been known as a powerful estrogen since more than fifty

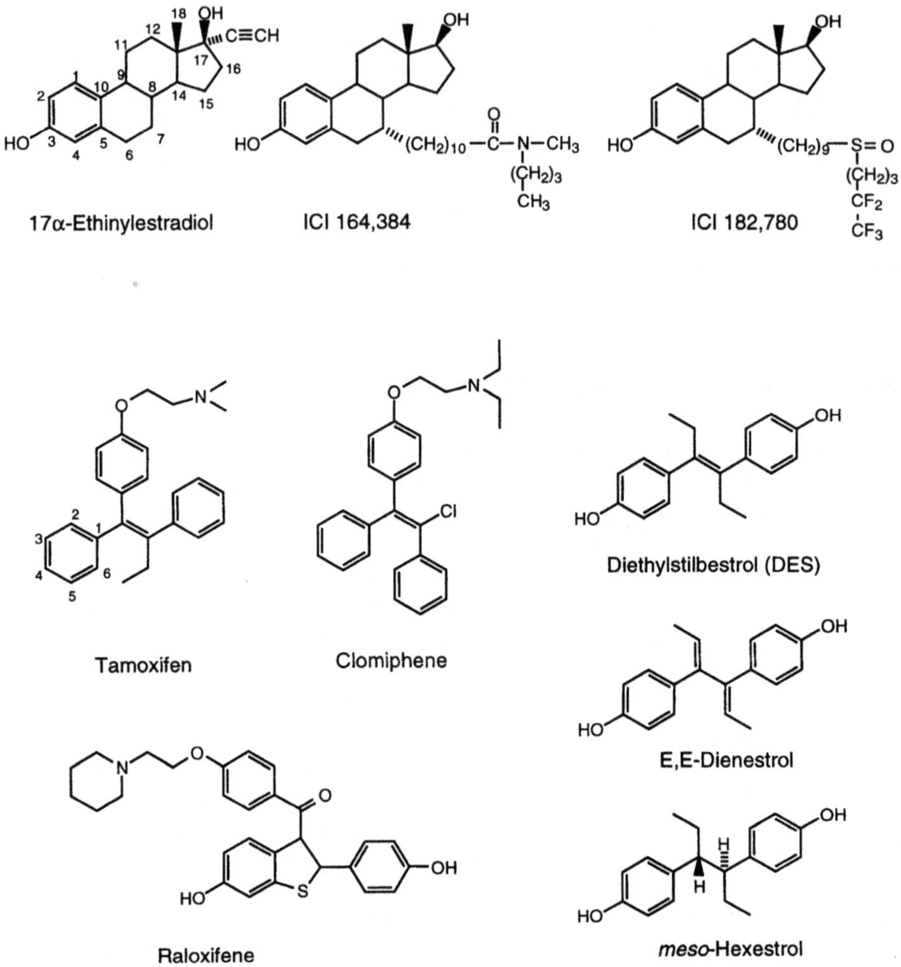

Fig. 9. Chemical structures of pharmaceutical estrogens and antiestrogens

years; it is superior to E2 with respect to oral efficacy and elimination half-life, and widely used directly or as its 3-*O*-methyl ether (mestranol) in oral contraceptives. The C7-alkyl-substituted E2 derivatives ICI 164,384 and ICI 182,780 are pure antiestrogens (see Chap. Mechanisms of Estrogen Receptor-Mediated Agonistic and Antagonistic Effects (Part I)) and often used in studies on the mechanisms of estrogen action. The triphenylethylene derivative tamoxifen is very popular in clinical practice for the treatment and prophylaxis of breast cancer, whereas clomiphene is often used for the induction of ovulation. 4-Hydroxytamoxifen, formed as a metabolite from tamoxifen, has an increased hormonal activity. Raloxifene is chemically related to the triphenylethylene-type compounds. The stilbene-type agents diethylstilbestrol (DES), *E,E*-dienestrol and *meso*-hexestrol were synthesized in the late 1930s and are among the first man-made estrogens used for human treatment [17]. Shortly after their introduction into human medicine, they were also employed on a large scale as growth promoters in livestock. Both of these uses are banned today due to their carcinogenicity for humans and several animal species.

3.1.2
Androgens and Antiandrogens

Typical examples for synthetic steroidal androgens are 17α-alkylated derivatives of testosterone, e.g., methyltestosterone (Fig. 10). This chemical modification of the testosterone molecule increases oral efficacy and elimination half-life. Elimination of the C19 methyl group decreases androgenic activity while maintaining anabolic activity. A compound related to 19-nortestosterone is 17β-trenbolone (Fig. 10), the acetate of which is used as growth promotor in cattle in several countries. An important antiandrogen with wide clinical use is cyproterone acetate (Fig. 10), which resembles synthetic progestins (see below) and has a pronounced progestagional activity. A synthetic antiandrogen without progestin-like side effects is the non-steroidal flutamide (structure not shown).

3.1.3
Progestins and Antiprogestins

Pharmaceutical progestins are either derivatives of 17α-hydroxyprogesterone or of 17α-ethinylated testosterone or 19-nortestosterone. Typical representatives of the first category are medroxyprogesterone acetate, megestrol acetate, and melengestrol acetate (Fig. 11). Melengestrol acetate is used to promote cattle growth in the USA and other countries. The second class of man-made progestins comprises ethisterone and norethisterone (Fig. 11). An important antiprogestational compound is mifipristone, better known as RU 486 and used for the interruption of early pregnancy.

3.2
Industrial Chemicals

As noted above, numerous chemicals produced for diverse industrial purposes exhibit sex hormone-like activities. Such compounds have been preferentially

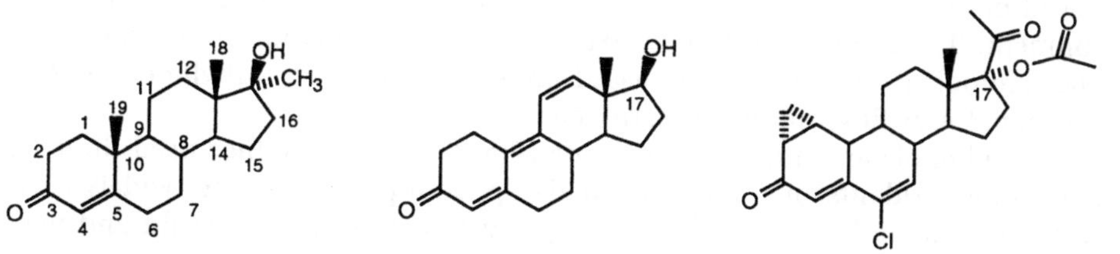

Fig. 10. Chemical structures of pharmaceutical androgens and antiandrogens

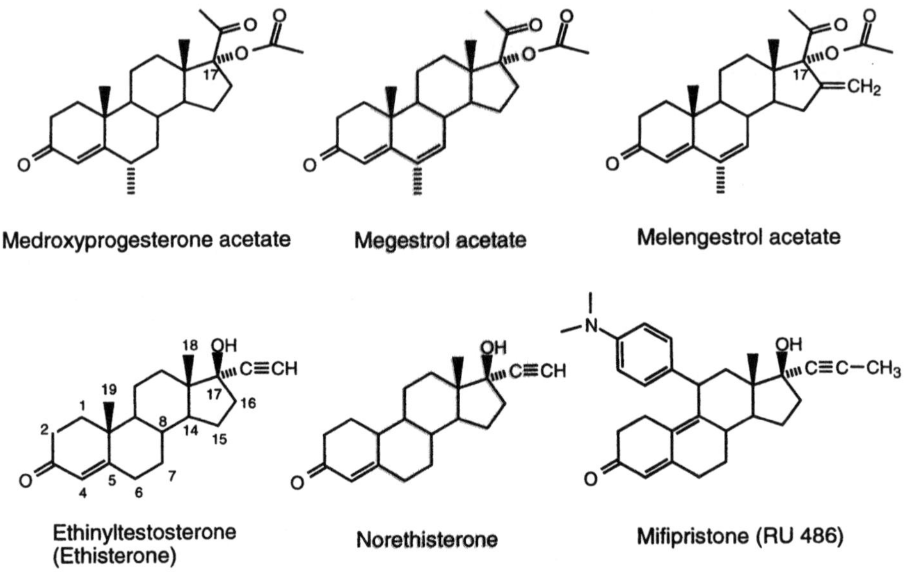

Fig. 11. Chemical structures of pharmaceutical progestins and antiprogestins

found in certain chemical classes, e.g., phenols, halogenated substances and phthalates. Typical representatives will be discussed below. It can be expected that the number of substances with hormonal activity and the number of chemical categories will increase in the future as more industrial chemicals will be tested.

3.2.1
Phenolic Compounds

The phenolic A ring of steroidal estrogens has long been considered a prerequisite for estrogenicity. The phenolic hydroxy groups are also of paramount importance for the high estrogenic activity of DES and other stilbene-type compounds. In early studies aiming to synthesize powerful estrogens, it has been observed that numerous other phenols exhibit hormonal activity [18].

Among the numerous phenolic compounds produced by the chemical industry, two classes are presently in the focus of scientific and public interest as potential endocrine disruptors, viz., alkylphenols and bisphenols. Both are produced in large quantities, and substantial amounts are released into the environment. Their production, use, exposure, bioaccumulation, biodegradation and biological effects are discussed in detail in Chap. Alkylphenols and Bisphenol A as Environmental Estrogens (Part I).

Alkylphenols are simple compounds (Fig. 12) that are either used directly, e.g., as antioxidants, or released from alkylphenol polyethoxylates widely used in the plastic and textile industry, in agriculture and for household and personal care items.

The prototype of bisphenols is bisphenol A (BPA, Fig. 12), used in large amounts for the production of polycarbonate plastics and epoxy resins (see

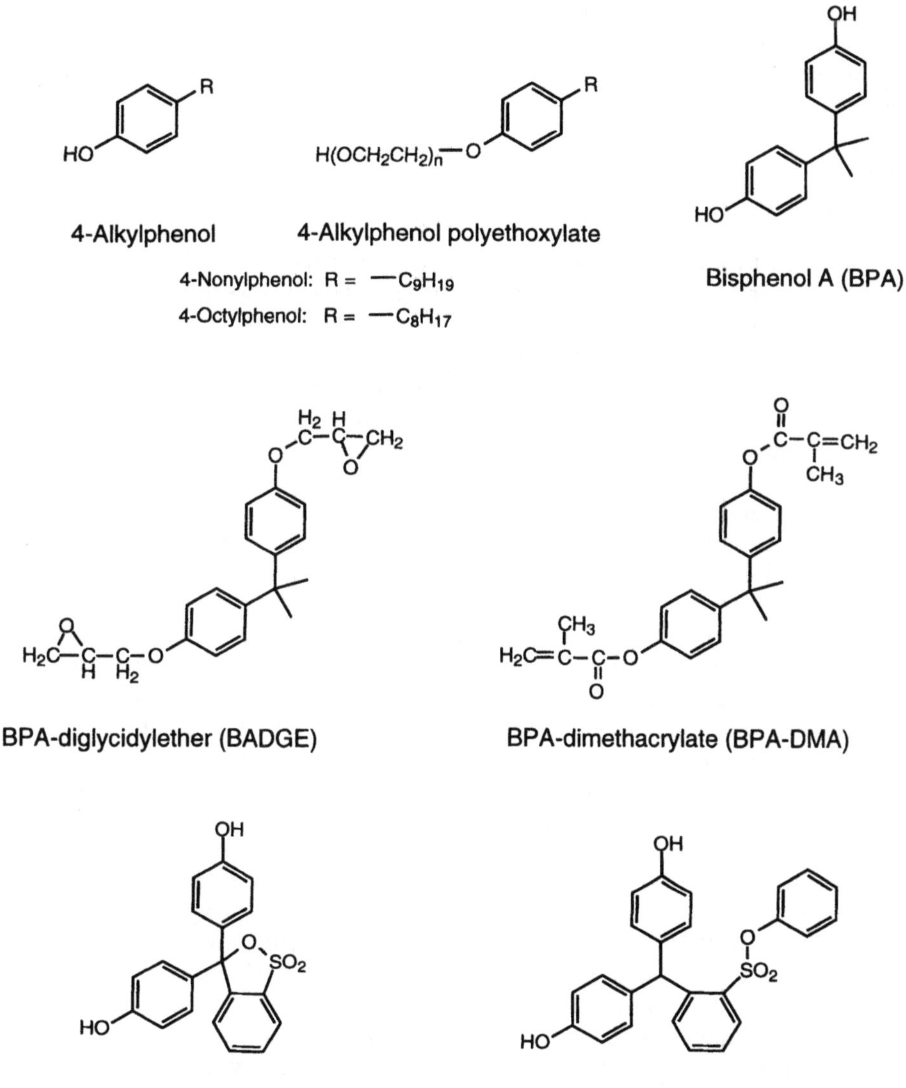

Fig. 12. Chemical structures of phenolic industrial chemicals

Chap. Alkylphenols and Bisphenol A as Environmental Estrogens (Part I)). For the latter, derivatives of BPA such as the diglycidyl ether (BADGE) and the dimethacrylate (BPA-DMA) are important. Their chemical structures are depicted in Fig. 12. Other bisphenols closely related to BPA carry different substituents at the benzylic carbon atom or additional groups at the phenolic rings.

Another bisphenol is phenol red (Fig. 12), which is a pH indicator commonly used in cell culture media. It was first reported that phenol red stimulates the growth of estrogen-responsive human breast cancer (MCF-7) cells and binds to the estrogen receptor in these cells [19]. Subsequent studies revealed that the estrogenicity is due to a minor impurity, identified as bis(4-hydroxyphenyl)-[2-(phenoxysulfonyl)phenyl]methane (Fig. 12), in the commercial preparations of phenol red [20].

3.2.2
Polychlorinated Compounds

It has been known since the 1950s that the then widely used insecticide 2,2-bis(*p*-chlorophenyl)-1,1,1-trichloroethane (*p,p′*-DDT) is weakly estrogenic (see Chap. Antiandrogenic Effects of Environmental Endocrine Disruptors (Part I)). The chemical structure of *p,p′*-DDT (Fig. 13) resembles that of bisphenols in which the phenolic groups are replaced by chlorine atoms. The isomeric *o,p′*-DDT (Fig. 13), which is present as an impurity in commercial *p,p′*-DDT, also exhibited estrogenic activity. 2,2-Bis(*p*-chlorophenyl)-1,1-dichloroethene (*p,p′*-DDE, Fig. 13), a major metabolite of *p,p′*-DDT, has recently been demonstrated to exhibit significant antiandrogenic activity (see Chap. Antiandrogenic Effects of Environmental Endocrine Disruptors (Part I)). The estrogenicity of methoxychlor (Fig. 13) is probably accounted for by a metabolite formed by oxidative demethylation; this metabolite, 2,2-bis(*p*-hydroxyphenyl)-1,1,1-trichloroethane (HPTE, Fig. 13), contains a bisphenol structure.

Other chlorinated pesticides, e.g., dieldrin, endosulfan and chlordecone (kepone, Fig. 13) were also disclosed as being weakly estrogenic in a systematic study screening environmental pollutants for their ability to elicit cell proliferation and other estrogenic effects in cultured human breast cancer (MCF-7) cells [21].

Polychlorinated dibenzo(*p*)dioxins (PCDD), polychlorinated dibenzofurans (PCDF) and polychlorinated biphenyls (Fig. 13) are dealt with in Chaps. Hydroxylated Polychlorinated Biphenyls (PCBs) and Organochlorine Pesticides as Potential Endocrine Disruptors (Part I) and Alterations in Male Reproductive Development: The Role of Endocrine Disrupting Chemicals (Part II). PCDD and PCDF are ubiquitous byproducts of the combustion of organic materials, in particular of plastics containing chlorine. The most toxic member of this class of compounds is TCDD.

Polychlorinated biphenyls (PCB) are among the most persistent and ubiquitous environmental pollutants. Whereas the PCB themselves have no or at best marginal estrogenicity, significant hormonal activity may be entailed to these molecules by hydroxylation [22]. For example, 4-hydroxy-2′,4′,6′-trichlorobiphenyl (Fig. 13) was about 0.01% as estrogenic as E2 in cultured MCF-7 cells [21]. See Chap. Hydroxylated Polychlorinated Biphenyls (PCBs) and Organochlorine Pesticides as Potential Endocrine Disruptors (Part I) for a detailed account of hydroxylated PCBs.

3.2.3
Phthalates

Another class of widely used industrial chemicals comprises the diesters of phthalic acid with various aliphatic alcohols. Worldwide, about sixty different

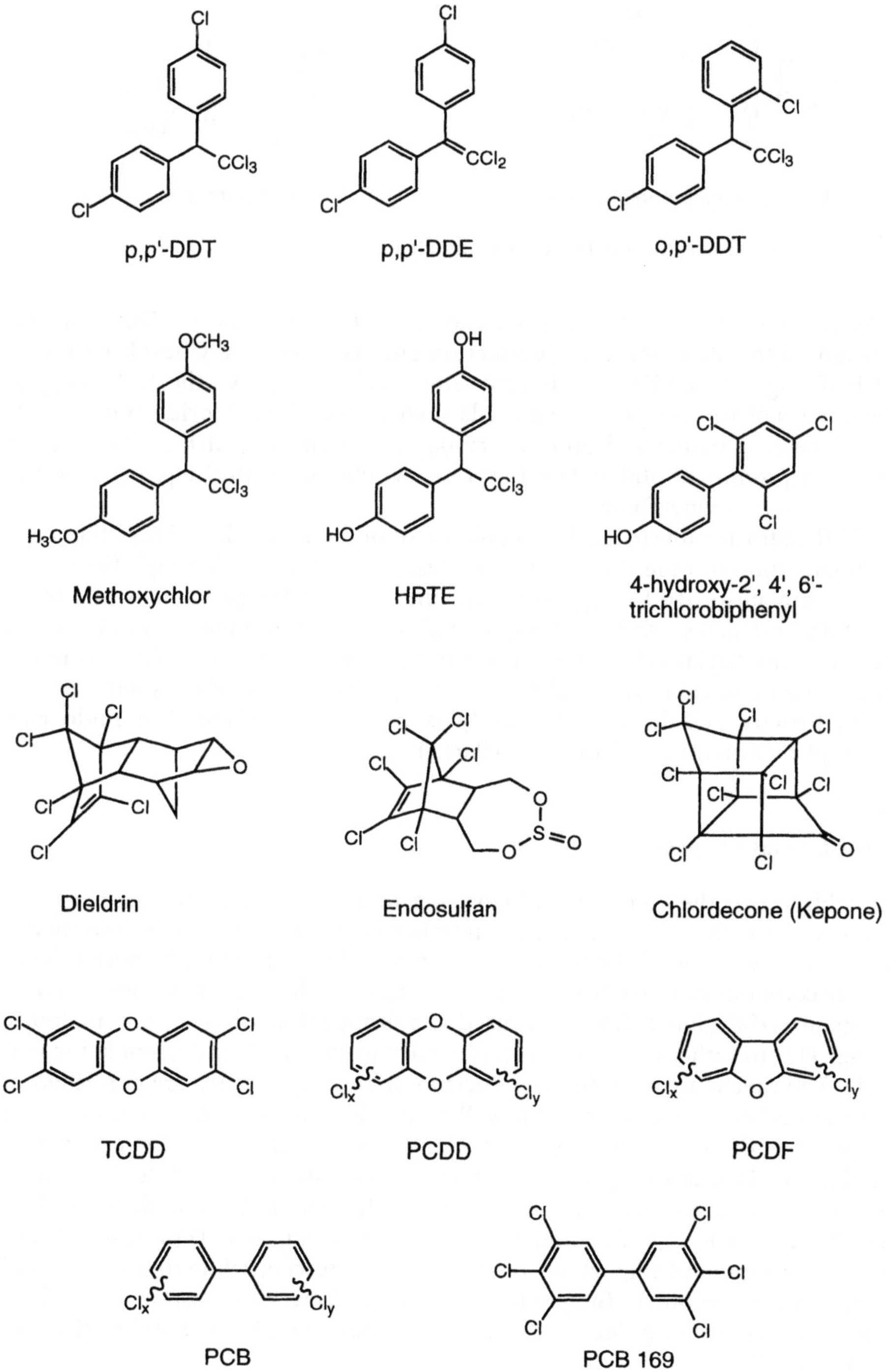

Fig. 13. Chemical structures of polychlorinated chemicals

Fig. 14. Chemical structures of two typical phthalates

phthalates are produced and applied for very diverse purposes. The most important phthalate in terms of production and use is di(2-ethylhexyl) phthalate (DEHP, Fig. 14). DEHP is primarily employed to improve the technological properties of various plastic materials such as polyvinyl chloride, but cosmetics, adhesives, paints and other every-day-chemicals may also contain DEHP. Another prominent phthalate is butyl benzyl phthalate (BBP, Fig. 14), which is mainly used in vinyl floor tiles.

Phthalates are able to leach from plastic or other materials and may thus contaminate food or other biological samples. Due to their high lipophilicity, they are readily absorbed through the skin, lung or from the gastrointestinal tract. Certain phthalates exhibit developmental toxicity on the male genital tract in rodents and display estrogenic activity in *in vitro* systems. A detailed account of the human and environmental exposure to phthalates, as well as their metabolism, toxicity and hormonal activity is provided in Chap. The Endocrine Disrupting Potential of Phthalates (Part I).

3.2.4
Miscellaneous Compounds

In addition to the compounds belonging to certain chemical classes as discussed above, there is an increasing number of substances with diverse chemical structures, which exhibit endocrine activity through multiple mechanisms. These compounds range from simple structures such as 1,2-dibromo-3-chloropropane (DBCP, Fig. 15), which is used as a fumigant to kill insects, nematodes, fungi, etc., to rather complex molecules such as the antifungal agent ketoconazole. The mechanisms of their endocrine disrupting activity are described in detail in other chapters of this book. Vinclozolin, procymidone and linuron are fungicides discussed in Chaps. Antiandrogenic Effects of Environmental Endocrine Disruptors (Part I) and Emerging Issues Related to Endocrine Disrupting Chemical and Environmental Androgens and Antiandrogens (Part II). Amsonic acid (4,4′-diaminostilbene-2,2′-disulfonic acid, DAS) is a DES-like stilbene compound important for dyes and fluorescent whitening agents, and fenarimol is another fungicide (see Chap. Emerging Issues Related to Endocrine Disrupting Chemical and Environmental Androgens and Antiandrogens (Part II).

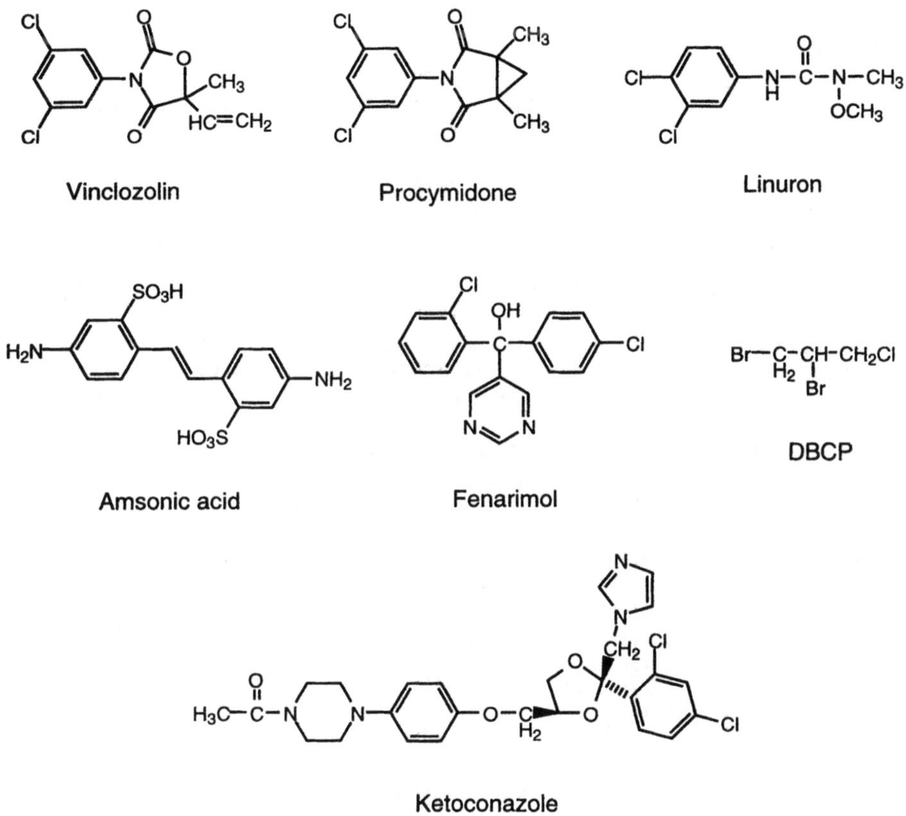

Fig. 15. Endocrine active compounds with diverse structures

4 Conclusion

This paper focuses on the chemical structures of typical compounds known today to exert sex-hormone-like activity. The diversity of structures is surprising, in particular for agents acting as estrogens or antiestrogens. As scientific and public interest in EAC prevails, and assays for their detection are improved and widely used, the list of such compounds will undoubtedly grow over the coming years. Moreover, the number of EAC must be expected to further increase when the metabolites of natural and anthropogenic chemicals are taken into account. Most *in vitro* assays for hormonal activity do not have the ability to metabolize the test compound. Therefore, chemicals needing metabolic alteration to unfold their hormonal activity must be expected to give negative results in such *in vitro* assays. Conversely, agents that are hormonally active *per se* may lose their activity upon biotransformation. Thus, not only the activity of the parent substance but also of its metabolites must be known before the potential of a compound to affect the endocrine system can be assessed.

5
References

1. Conn PM, Melmed S (eds) (1997) Endocrinology: Basic and Clinical Principles. Humana Press, Totawa, New Jersey, USA
2. National Research Council (US) (1999) Hormonally Active Agents in the Environment. National Academy Press, Washington DC, USA
3. Price KR, Fenwick GR (1985) Food Addit Contam 2:73
4. Farnsworth NR, Bingel AS, Cordell GA, Crane FA, Fong HHS (1975) J Pharmacol Sci 64:717
5. Stitch RS, Toumba JK, Groen MB, Funke CW, Leemhuis J, Vink J, Woods GF (1980) Nature 287:738
6. Setchell KDR, Axelson M, Sjövall J, Gustafsson BE (1982) Nature 298:659
7. Meagher LP, Beecher GR, Flanagan VP, Li BW (1999) J Agric Food Chem 47:3173
8. Obermeyer WR, Musser SM, Betz JM, Casey RE, Pohland AE, Page SW (1995) Proc Soc Exp Biol Med 208:6
9. Umezawa T, Davin LB, Lewis NG (1991) J Biol Chem 266X:10210
10. Breinholt V, Larsen JC (1998) Chem Res Toxicol 11:622
11. Milligan SR, Kalita JC, Heyerick A, Rong H, De Cooman L, De Keukeleire D (1999) J Clin Endocrinol Metab 84:2249
12. Ritchey JG, Waterhouse AL (1999) Am J Enol Vitic 50:91
13. Frankel EN, Waterhouse AL, Teissedre PL (1995) J Agric Food Chem 43:890
14. Frankel EN, Waterhouse AL, Kinsella J (1993) Lancet 341:1103
15. Ribeiro de Lima MT, Waffo-Teguo P, Teissedre PL, Pujolas A, Vercauteren J, Cabanis JC, Merillon JM (1999) J Agric Food Chem 47:2666
16. Kennedy DG, McEvoy JDG, Blanchflower WJ, Hewitt SA, Cannavan A, McCaughey WJ, Elliot CT (1998) J Med Vet Ser B 42:509
17. Smith OW (1948) Am J Obstet Gynecol 56:821
18. Solmssen UV (1945) Chem Rev 36:481
19. Berthois Y, Katzenellenbogen JA, Katzenellenbogen BS (1986) Proc Natl Acad Sci USA 83:2496
20. Bindal RD, Katzenellenbogen JA (1988) J Med Chem 31:1978
21. Soto AM, Sonnenschein C, Chung KL, Fernandez MF, Olea N, Olea Serrano F (1995) Environ Health Perspect 103 (Suppl 7):113
22. Korach KS, Sarver P, Chae K, McLachlan JA, McKinney JD (1988) Mol Pharmacol 33:120

CHAPTER 5

Exposure to Endogenous Estrogens During Lifetime

Jörg Dötsch[1], Helmuth G. Dörr[1], Ludwig Wildt[2]

[1] Klinik für Kinder und Jugendliche, Friedrich-Alexander-University Erlangen-Nürnberg, Loschgestrasse 15, 91054 Erlangen, Germany
e-mail: JoergWDoetsch@yahoo.com
[2] Universitätsfrauenklinik, Friedrich-Alexander-University Erlangen-Nürnberg, Loschgestrasse 15, 91054 Erlangen, Germany

The present review summarizes data on the time course and physiological function of the three major endogenous estrogens estrone (E1), 17β-estradiol (E2), and estriol (E3) during the different phases of life in the human female and male. During fetal life, E3 is the most abundant estrogen produced by the fetoplacental unit. E3 affects cerebral development, leads to breast gland swelling in both girls and boys and promotes uterine growth up to a size that is not reached again until puberty. In infancy and childhood estrogen levels are low before the ovaries are stimulated to increase the production of E2 at puberty. In the complex course of maturation, the onset of puberty is characterized by a gradually increasing pulsatile secretion of hypothalamic gonadotropin-releasing hormone followed by a gradual rise of circulating gonadotropin levels. Increasing E2 concentrations in girls promote development of female sex characteristics, menarche, behavioral changes, pubertal growth spurt and finally the closure of epiphysal growth zones. Throughout fertile life of the human female, ovarian E2 remains the major endogenous estrogen. It is produced by the granulosa cells of the growing follicle as well as by the corpus luteum. Among other functions, it is important for endometrial proliferation, as a prerequisite for blastocyst implantation and pregnancy. E2 induces growth of the uterus and maturation of the breast. E2 production declines gradually during late reproductive life; as a consequence, menstrual bleeding ceases with menopause. During postmenopause, the predominant endogenous estrogen is E1, which is mainly produced by adipose tissue from androgenic precursors secreted by the ovarian stroma and the adrenal gland. Decreased estrogen concentrations lead to atrophy of the inner and outer genitalia, osteoporosis, an increased risk of cardiovascular disease, hot flashes and emotional instability.

In summary, exposure to endogenous estrogens during lifetime in the female varies by several orders of magnitude. The time course of estrogen concentration is characterized by a high-estrogen environment during pregnancy, a decline following birth to the low levels during prepuberty. Onset of sexual maturation is indicated by the rising levels of E2 reaching adult concentrations some years after menarche.

Keywords. Estradiol, Estriol, Estrone, Menopause, Puberty

1 **Introduction and Physiology of Estrogen Metabolism** 83

1.1 Biochemistry of Endogenous Estrogens 83
1.2 Action of Estrogens in the Target Cell 85
1.3 Physiology of Estrogen Action 85

The Handbook of Environmental Chemistry Vol. 3, Part L
Endocrine Disruptors, Part I
(ed. by M. Metzler)

2 Estrogens in Fetal Life, Infancy and Prepubertal Childhood . . . 86

2.1 Fetus and Neonate . . . 86
2.2 Infancy and Childhood . . . 86

3 Role of Estrogens During Puberty . . . 87

3.1 Female Puberty . . . 87
3.2 Male Puberty . . . 88

4 Role of Estrogens in the Adult Woman and Man During Reproductive Age . . . 88

4.1 Estrogens During Menstrual Cycle Regulation of the Ovarian and Uterine Function . . . 88
4.2 Estrogen Regulation of the Mucosal Immune System . . . 89
4.3 Estrogen Effects Outside the Reproductive System . . . 89
4.4 Adverse Effects of Endogenous Estrogens . . . 89
4.5 Role of Endogenous Estrogens for the Fertile Man . . . 90

5 Role of Estrogens in Conception and Pregnancy . . . 90

5.1 Conception and Blastocyst Implantation . . . 90
5.2 Estrogen Production in Pregnancy . . . 90
5.3 Impact of Estrogens on the Mother During Pregnancy . . . 92
5.4 Role of Estrogens for Lactation . . . 93

6 Menopause and Postmenopause . . . 93

6.1 Physiology and Estrogen Concentrations . . . 93
6.2 Consequences of Decreasing Estrogen Concentrations During Menopause . . . 94

7 Tables of Concentrations of Estrogens During Different Phases of Life . . . 95

8 References . . . 98

Abbreviations

cAMP	cyclic adenosine monophosphate
DHEA	dehydroepiandrosterone
DHEAS	dehydroepiandrosterone sulfate
E1	estrone
E2	17β-estradiol
E3	estriol
ERα	estrogen receptor alpha
ERβ	estrogen receptor beta
FSH	follicle-stimulating hormone

GnRH	gonadotropin-releasing hormone
16α-HO-DHEA	16α-hydroxydehydroepiandrosterone
16α-HO-DHEAS	16α-hydroxydehydroepiandrosterone sulfate
IgA	immunoglobulin A
IgG	immunoglobulin G
LH	luteinizing hormone
SHBG	sex hormone-binding globulin

1 Introduction and Physiology of Estrogen Metabolism

1.1 Biochemistry of Endogenous Estrogens

The most important endogenous estrogens in humans are 17β-estradiol (E2), estrone (E1) and estriol (E3, Fig. 1). They all are steroids consisting of 18 carbon atoms and characterized by an aromatic A ring. For the specific estrogen effect the aromatic A ring and hydroxy groups at positions 3 and 17 are indispensable.

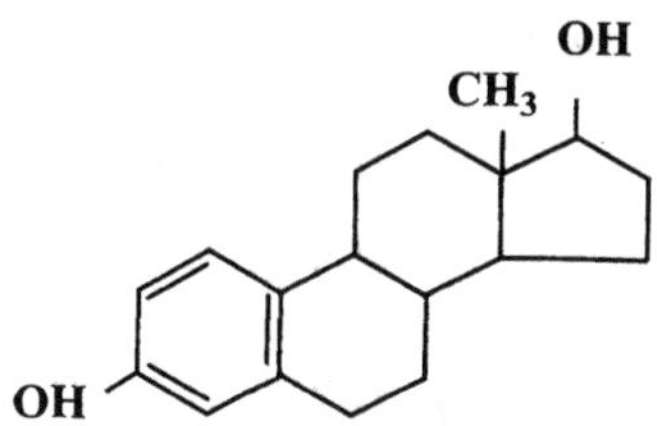

Estra*di*ol (E2)

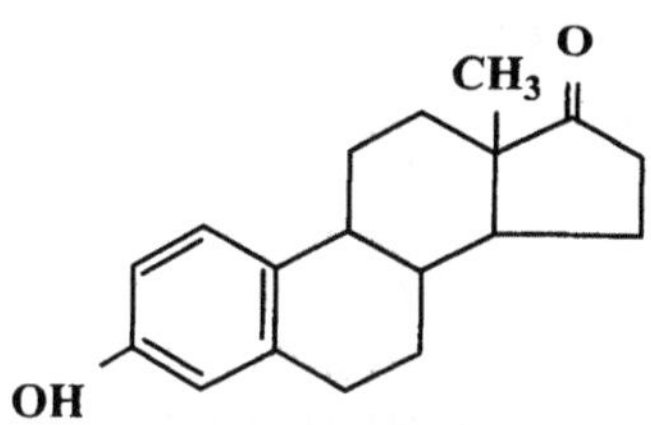

Estrone (E1)

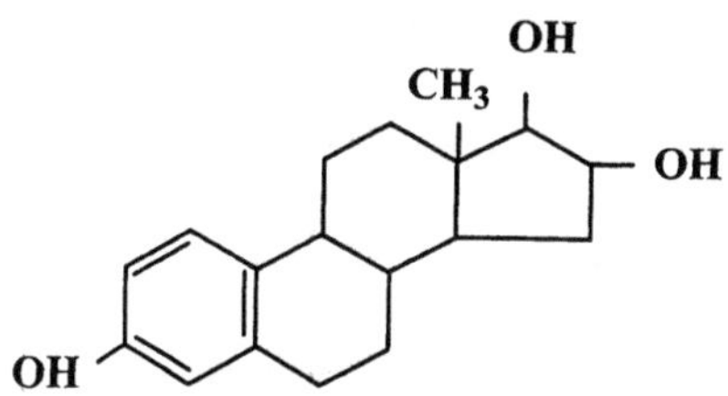

Es*tri*ol (E3)

Fig. 1. Chemical structures of the three most important endogenous estrogens estradiol, and estrone estriol

E2, the most potent and important estrogen in non-pregnant women, is predominantly produced by the granulosa cells of the active follicle from androgens delivered by the theca interna (two-cell hypothesis). During pregnancy, E3 produced from androgenic precursors provided by the fetus and the mother, respectively, represents the major estrogen [1]. E1, the third of the major endogenous estrogens, exists in metabolic equilibrium with E2 due to the action of 17β-hydroxysteroid dehydrogenase. In the classic pathway, the estrogen synthesis starts from cholesterol provided by lipoproteins (Fig. 2).

Estrogens are biologically inactivated and excreted after sulfation or glucuronidation, respectively, allowing renal excretion of the inactivated steroids. Although considerable amounts of conjugated estrogens are excreted into the bile, only a small fraction appears in the feces. The majority of the conjugates is reabsorbed after hydrolysis by bacteria from the gastrointestinal tract (enterohepatic circulation).

The majority of E2 (98%) circulates bound to albumin or to sex hormone-binding globulin (SHBG), a specific carrier protein that binds estrogens and androgens with high affinity.

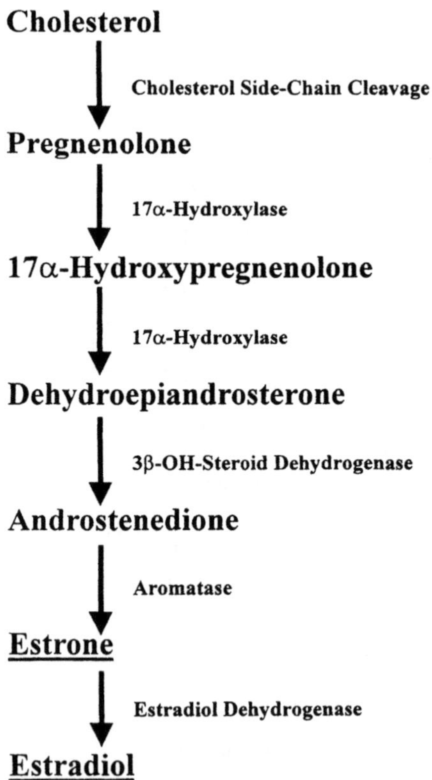

Fig. 2. Enzymatic steps in the classical pathway of estradiol biosynthesis in the ovary

1.2 Action of Estrogens in the Target Cell

Estrogens can enter their target cells via passive diffusion through the cell membrane. There is, however, increasing evidence for an active transport mediated by a specific importer localized within the plasma membrane. After transport through the cell membrane, estrogens bind to specific receptors located within the nucleus of the target cells [2, 3]. There are two different receptors for E2: estrogen receptor alpha (ERα) and estrogen receptor beta (ERβ) that can form heterodimers exhibiting different affinities to specific DNA sequences termed estrogen response elements [4–7]. As evidenced by knockout mice experiments, the ERα appears to be more important for gene transcription in the target tissues. ERβ knockout mice are less severely affected resulting predominantly in subfertility [4]. In conjunction with a number of local growth factors, estrogens stimulate growth and differentiation of tissues and organs. See Chap. 1 for a detailed description of estrogen action in target cells.

The estrogenic activity of E1, E2 and E3 varies considerably when expressed on a molar basis. While E2 represents the most active compound, exerting all estrogenic effects, the biological activity of E3 and E1 appears to be much lower.

E2 induces the transcription of its own receptors and stimulates the biosynthesis of progesterone receptors as a prerequisite of progesterone action. Conversely, progesterone and other progestins inhibit the transcription of estrogen receptors. Therefore, progestins exhibit an anti-estrogenic effect.

1.3 Physiology of Estrogen Action

Estrogens are responsible for the emergence of secondary sex characteristics. These include breast development, typical female body proportions, distribution of subcutaneous adipose tissue and the characteristic estrogen-dependent changes of the female genital tract. Target organs for the estrogens are the external genitalia, vagina, uterus, fallopian tubes and the ovaries. Target tissues outside the reproductive organs are, among others, skin including its annexes, bones, the cardiovascular system, the central nervous system, and the liver.

In the reproductive organs, endogenous estrogens promote cell proliferation. Blood flow, water retention, and the accumulation of amino acids and proteins are increased. The uterine cervix is stimulated by estrogens to secrete mucus. Estrogens stimulate alveolar growth of the breast; however, for full maturation of the mammary gland the additional action of progestins, prolactin as well as glucocorticoids and insulin are necessary.

2 Estrogens in Fetal Life, Infancy and Prepubertal Childhood

2.1 Fetus and Neonate

The fetus is exposed to very high estrogen concentrations due to the enormous E3 production by the placenta (see below). Estrogens in fetal life seem to be important for the development of a number of functions, including brain development [8, 9]. Newborn female as well as male knockout mice for ERα have severe abnormalities of the reproductive tract: female mice exhibit a hypoplastic uterus and hemorrhagic ovaries [4, 10], whereas male mice show testicular dysgenesis. Both genders are infertile. On the other hand, ERβ knockout mice have a normal phenotype and only female mice are subfertile. The knockouts of both receptors show an even more severe phenotype than the ERα knockouts.

However, in humans estrogens do not appear to be essential for fetal survival, placental growth, or male and female sexual differentiation. This is concluded from studies on estrogen deficiency due to mutations in the aromatase gene and estrogen resistance due to disruptive mutations in the estrogen receptor gene [11]. Also sulfatase deficiency does not lead to an abnormal fetal development.

Quite frequently, human newborns of both sexes exhibit hypertrophy of the breast glands. Since lactation is initiated by a drop of circulating estrogens, the sudden decline of serum estrogen levels after clamping the umbilical cord may even lead to temporary mammary secretion called "witch's milk". In newborn female babies the uterine size is much larger than that of older infants and involutes to about one third of its birth size during the following 6–12 months. Intrauterine endometrial proliferation induced by estrogens may result in withdrawal bleeding in the female newborn [12]. The vaginal epithelium right after birth resembles that of the fertile woman, exhibiting low pH and glycogen storage. After the first week of life there is a significant rise of gonadotropin concentrations due to the drop of estrogen levels and the loss of the feedback inhibition on pituitary gonadotropin secretion. As a consequence, pulsatile secretion of gonadotropin-releasing hormone (GnRH) promotes the release of follicle-stimulating hormone (FSH), resulting in follicular growth in the newborn girl's ovary as may be demonstrated by ultrasonography. On the average, E2 levels in newborn girls are slightly higher than in boys as a consequence of the increased ovarian activity.

2.2 Infancy and Childhood

Until the end of the second year of life the pulsatile GnRH secretion continuously diminishes, resulting in a decline of pituitary gonadotropin secretion and a reduction of ovarian stimulation. In toddlers, GnRH release is severely reduced as indicated by the low-amplitude, low-frequency peaks of luteinizing hormone (LH) that occur every three to four hours during sleep [13]. E2 concentrations are usually at or below the detection limit, although newer, more sensitive assays for the detection of E2 show certain fluctuations during that period of life. Using an ultrasensitive assay, elevated E2 concentrations can be detected in girls with idiopathic premature telarche [14].

Between the sixth and ninth year of life, the hypothalamic activity increases slowly as reflected by the increase of the number and amplitude of FSH and LH pulses. Via a G-protein-coupled receptor, FSH increases intracellular cyclic adenosine monophosphate (cAMP) concentrations leading to an induction of the aromatase that is essential for estrogen synthesis. However, FSH and estrogen levels still remain low until the onset of puberty.

3 Role of Estrogens During Puberty

3.1 Female Puberty

In female infancy and childhood, the activity of the hypothalamic pulse generator that directs GnRH release is almost completely inhibited and LH and FSH serum levels are low. Small increases of serum E2 are sufficient to further inhibit pituitary gonadotropin secretion. The mechanisms inhibiting hypothalamic GnRH release are not known at present [15, 16] nor are the mechanisms resulting in the release of inhibition during puberty [17]. From studies in ovariectomized primates it is known that estrogens are not decisive for the activation of the GnRH pulse generator but rather have modulating effects on gonadotropin release [18]. As a consequence of ovarian stimulation by gonadotropins, E2 levels in serum increase. In conjunction with adrenal as well as ovarian androgens, pubic hair develops. Ovaries, uterus, the fallopian tube, vagina, and breast glands grow under the influence of E2. In puberty stage 3–4 according to Tanner, when E2 levels of approximately 40 pg/mL (140 nmol/L) are reached, the endometrium has sufficiently proliferated to allow withdrawal bleedings after gestagen administration. However, the first menstrual bleeding, the menarche, is usually a result of a temporary slight drop in E2 concentrations leading to an estrogen withdrawal hemorrhage. In Europe, menarche occurs around a mean age of 13.4 (11–15.6) years. Interestingly, 150 years ago the mean age of menarche was 17 years [19, 20].

While there is no doubt that estrogens as well as androgens have profound effects on behavior, controversy persists as to whether these effects are permanent (imprinting) or not [21–23]. The classic example of brain imprinting by sex steroids, the androgen-induced permanent inactivation of the LH surge mechanism described in the laboratory rodent, does not seem to occur in primates or in humans [24, 25].

One important role of endogenous E2 is the promotion of growth and epiphysial closure to induce growth arrest. Estrogen deficiency due to mutations in the aromatase gene and estrogen resistance due to disruptive mutations in the estrogen receptor gene lead to absence of the pubertal growth spurt, delayed bone maturation, unfused epiphyses, continued growth into adulthood and very tall adult stature in both sexes [11].

3.2 Male Puberty

In about 60% of healthy pubertal boys there is clinically visible development of breast tissue that is called gynecomastia of puberty. Although these boys have normal serum concentrations of E2, testosterone, LH and FSH, an increased ratio of E2 to testosterone has been reported [26]. However, this observation has not been confirmed in more recent studies. Other investigators have observed elevated E2 concentrations in the early morning hours. Breast swelling in boys may also be explained by the combined effects of an androgen-induced decline in hepatic SHBG synthesis and the higher affinity of SHBG for testosterone as compared to E2, resulting in a shift of the balance between androgens and estrogens in favor of estrogenic hormones as long as androgen levels in plasma have not yet reached final adult levels [27, 28].

Estrogens increase the number of prolactin receptors in breast tissue. Also, E2 has a direct effect on the growth of mammary ducts. In some boys there is a familial increase in the extraglandular aromatization of C_{19}-steroids raising serum E1 concentrations and increasing the incidence of gynecomastia [29]. In addition to the extraglandular estrogen synthesis, e.g., in adipose tissue, E2 is produced in testicular tissue leading to E2 concentrations in the pubertal boy that are much higher than those found before puberty.

Estrogen deficiency due to mutations in the aromatase gene and estrogen resistance due to disruptive mutations in the estrogen receptor gene have no effect on normal male sexual maturation in puberty [11].

4 Role of Estrogens in the Adult Woman and Man During Reproductive Age

4.1 Estrogens During the Menstrual Cycle Regulation of the Ovarian and Uterine Function

Increasing concentrations of FSH induce the aromatization of androgens in the granulosa cells of the ovary, thus elevating E2 concentrations. E2 and FSH increase the FSH receptor concentration of the granulosa cells of the ovarian follicle. The peripheral E2 concentrations increase further and lead, together with ovarian inhibin, to a feedback inhibition of FSH secretion. When E2 levels exceed a certain threshold for a defined period of time, indicating the full maturation of the ovarian follicle, a massive increase of pituitary LH and FSH secretion is induced resulting in ovulation and corpus luteum formation. However, the growth of preovulatory follicles can proceed with minimal concentrations of LH and FSH in the presence of low peripheral estrogen levels [30]. Oocyte maturation and fertilization may proceed independently of ambient estrogen levels. This leads to the assumption that estrogens exert a minimal autocrine-paracrine function [31].

The rising E2 levels in the follicular phase result in proliferation of the uterine endometrium and in an increase of the number of glands. There is an increase in the amount and a change in the physicochemical properties of the cervical mucus (Spinnbarkeit). The decline of E2 and progesterone in the late luteal phase leads to a loss of endometrial blood supply and eventually to the onset of menses.

4.2 Estrogen Regulation of the Mucosal Immune System

The mucosal immune system in the female reproductive tract is the first line of defense against pathogenic organisms. Immunoglobulin A (IgA) and IgG levels in uterine secretions change markedly during the rat estrous cycle, with higher levels measured at ovulation than during any other stage of the cycle [32]. When ovariectomized animals are treated with E2, IgA and IgG levels markedly rise relative to untreated controls. These results underline the role of estrogens in the regulation of the local uterine defense mechanisms, enabling a pathogen-free environment for the implantation of the blastocyst.

4.3 Estrogen Effects Outside the Reproductive System

The protective effects of endogenous estrogens against disorders of the cardiovascular system, the skeletal system and central nervous functions have been thoroughly investigated with respect to the estrogen deficiency after menopause. Therefore, these aspects will be discussed in the section on the menopause. It should be stated, however, that young women with estrogen deficiency due to mutations in the aromatase gene and estrogen resistance due to disruptive mutations in the estrogen receptor gene achieve no normal bone mineral mass and have disturbances in insulin sensitivity and lipid homeostasis [11]. Similar observations were made in young women with amenorrhea.

4.4 Adverse Effects of Endogenous Estrogens

Despite the protective effects of estrogens on many organ systems in adult women, it is well known that some effects on the breast and the uterine endometrium are undesirable. Long-lasting uterine exposure to E2 alone results in endometrial hyperplasia and may ultimately promote endometrial cancer. The role of endogenous estrogens for the development of breast cancer is less clear. Estrogen receptor variants and mutations have been shown to be associated with a higher risk of breast cancer [33]. It could be demonstrated by a number of groups that blocking of the estrogen-receptor by the anti-estrogen tamoxifen lowers the recurrence of breast cancer or reduces the size of the tumor. Moreover, the additional production of estrogens by the adipose tissue may enhance the risk of breast cancer in premenopausal as well as postmenopausal women [34]. However, it is not clear to what extent these observations depend on changes in the metabolism of progestins.

4.5 Role of Endogenous Estrogens for the Fertile Man

The role of estrogens in the genesis of gynecomastia during puberty has already been discussed. However, elevated estrogen concentrations may also lead to gynecomastia in the adult man. A typical example is found in hepatic cirrhosis that results in a reduced hepatic ability to conjugate active estrogens and to render them inactive.

Endogenous estrogens are absolutely necessary for the reproductive capabilities in mice. Knockout experiments clearly show that male mice lacking the estrogen receptor alpha are infertile [10]. These mice exhibit atrophy of the testes and dysmorphogenesis of the seminiferous tubule. The prenatal development of the reproductive tract, however, is not affected by the estrogen receptor knockout. In human males, however, estrogen deficiency due to mutations in the aromatase gene and estrogen resistance due to disruptive mutations in the estrogen receptor gene is predominantly associated with reduced bone mass and delayed epiphysial closure [10, 11].

5 Role of Estrogens in Conception and Pregnancy

5.1 Conception and Blastocyst Implantation

In the ovary, follicles grow under the influence of gonadotropins and local growth factors. During the normal menstrual cycle, only one follicle of the cohort of growing follicles reaches the preovulatory stage while the remaining follicles undergo atresia. In humans, the E2 levels rapidly increase during the last five days before ovulation. After ovulation, the ruptured follicle luteinizes and the resulting corpus luteum secretes progesterone and E2. E2 secretion is increased during the luteal phase, corresponding to the increase in progesterone. The magnitude of the E2 secretion increase during the luteal phase, however, is not as high as during the follicular phase [35]. The time course of E2, progesterone, LH, FSH and prolactin secretion during the menstrual cycle is shown in Fig. 3.

After conception, E2 plays an important role for the implantation of the blastocyst in the uterus. It has been shown in ovariectomized rodents that the preexposition of the uterus to progesterone alone does not lead to a successful implantation of the blastocyst [36]. Implantation can be induced in ovariectomized mice and rats by doses of estrogens that are much smaller than those needed for behavioral changes [37]. The estrogen action is mediated by growth factors at the local sites [36].

5.2 Estrogen Production in Pregnancy

The regulation of estrogen production during pregnancy has been thoroughly investigated two to three decades ago. Prior to four weeks of gestation, the ma-

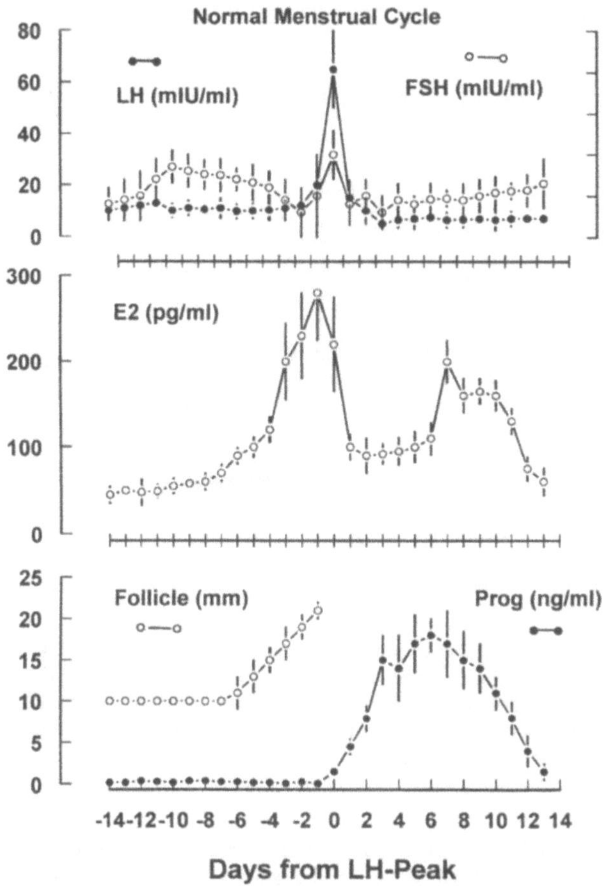

Fig. 3. Estradiol (E2) concentrations during the normal menstrual cycle in relation to progesterone (Prog), LH, FSH and follicle size

jority of E2 secreted into maternal blood is synthesized in the maternal ovaries [38]. After the fourth postovulatory week, however, bilateral oophororectomy or surgical removal of the corpus luteum does not diminish the levels of estrogens excreted in urine [39]. By seven weeks of gestation, the majority of estrogens entering both maternal and fetal compartments is of placental origin [40].

Near term, estrogen production is 1000 times the average daily estrogen production in normal ovulatory women. As a consequence, the maternal plasma concentration of E3, the most important estrogen in pregnancy, reaches levels at term that are 1000-fold those in the plasma of non-pregnant women [41]. Similarly, the maternal E2 and E1 concentrations in plasma increase from 50–100 pg/mL (on average) to 30,000 pg/mL at term [42]. Fifty percent of E2 near term is derived from fetal adrenal dehydroepiandrosterone sulfate (DHEAS) and 50% from maternal DHEAS (Fig. 4). On the other hand, 90% of E3 in maternal plasma is produced by the placenta from fetal plasma 16α-hy-

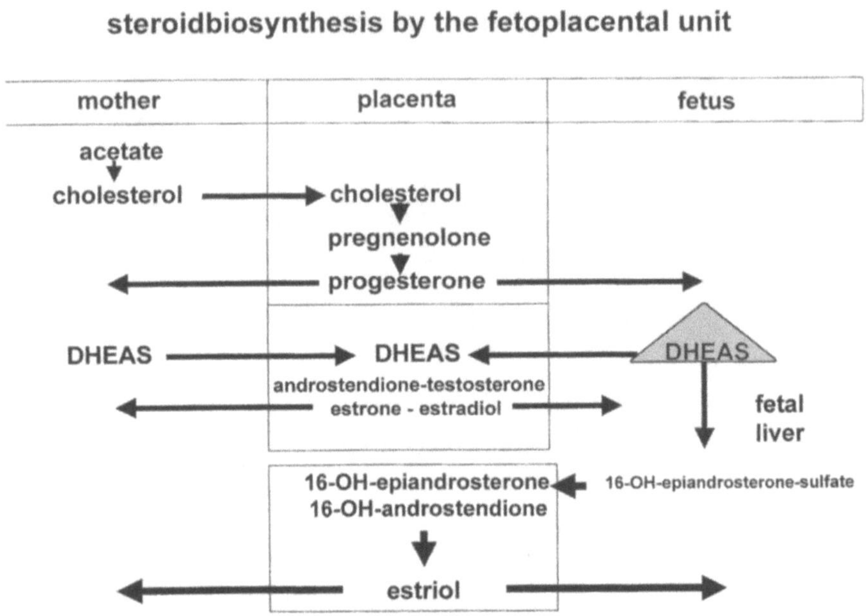

Fig. 4. Estrogen biosynthesis in the human placenta. Estriol, quantitatively the most important estrogen produced by the placenta, is synthesized from 16α-hydroxydehydroepiandrosterone sulfate (16α-HO-DHEAS), which is predominantly supplied by fetal blood. Estradiol is also produced by the normal placenta from dehydroepiandrosterone sulfate (DHEAS) provided by fetal and maternal blood [1]

droxydehydroepiandrosterone sulfate (16α-HO-DHEAS) and only 10% from all other sources. 16α-HO-DHEAS is produced by the fetal liver by sulfation of 16α-hydroxydehydroepiandrosterone (16α-DHEA), which is synthesized by the fetal adrenal gland. Due to the absence of 17α-hydroxylase (see Fig. 2), the *de novo* synthesis of estrogen from cholesterol or C_{21} steroids in human placenta is impossible [1]. 80–90% of the steroids produced in the placenta are secreted into the maternal blood [1]. Details of the synthesis of estrogens in the placenta and of the function of the fetoplacental unit are shown in Fig. 4.

5.3 Impact of Estrogens on the Mother During Pregnancy

The elevated estrogen concentrations during pregnancy induce the synthesis of transport proteins like thyroxin-binding globulin, transcortin and SHBG in the liver. Due to the high exposure to E3 in pregnancy, a number of tissues are stimulated to proliferate. In particular, uterine size increases up to 300-fold, an effect that is not observed in ERα knockout mice [10]. The development of the breast ducts is also highly stimulated. Together with a number of other hormone systems that are activated in pregnancy, E3 is the predominant estrogen in pregnancy leading to water and electrolyte retention and a number of other well known gestational changes. However, at least in the human female, E3 does not appear to be of critical importance for the maintenance of pregnancy, since sulfatase deficiency, a disorder resulting in extremely low E3 concentrations in maternal plasma, has no apparent adverse effect on the course of pregnancy.

5.4 Role of Estrogens for Lactation

During pregnancy, the role of the endogenous estrogens is to stimulate the growth of breast gland ducts [43], whereas the alveolar development is predominantly induced by progesterone. The high E3 concentrations later in pregnancy delay the actual onset of lactation. Therefore, only after removal of the placenta as the major source of estrogen synthesis, does the effect of prolactin on milk production become evident. Induction of milk production and galactorrhea have also been observed after betamethasone administration for prophylaxis of the respiratory distress syndrome of the neonate. Betamethasone is known to reduce E3 serum levels [44]. Similar effects on lactation are observed after a rapid decline of E3 serum levels related to fetal distress.

6 Menopause and Postmenopause

6.1 Physiology and Estrogen Concentrations

Over the past ten years, the menopause has probably been the most investigated field of research dealing with endogenous and exogenous estrogens [45]. Only the role of estrogens and antiestrogens in the treatment of breast cancer has been of similar interest. Menopause is defined as the last spontaneous menstruation. In Europe, the mean age of menopause is 51 years. In contrast to menarche, the age of menopause has not changed throughout the last century. Approximately five years prior to the absolute failure of ovarian hormone production, the first clinical indicators of disturbances of estrogen and progesterone production manifest with irregular menstrual bleedings. This phase is referred to as premenopause [46]. Whereas progesterone production drops relatively fast during that phase (Fig. 5), E2 synthesis decreases more gradually. These hormonal changes reflect the loss of ovarian follicles that may be stimulated. In addition, ovarian blood vessels show regressive changes and eventually obliterate.

With the progression of menopause, the E2 levels in the circulation decrease considerably until they reach concentrations that are frequently below 20 pg/mL. These concentrations are insufficient to induce adequate endometrial proliferation and subsequent menstrual bleeding. Ovariectomy in postmenopausal women does not lead to a further decrease in E2 concentrations, indicating the absolute loss of ovarian function [47]. Because the negative feedback on pituitary gonadotropin secretion is lost, there is a significant continuous increase in serum LH and FSH concentrations (Fig. 5).

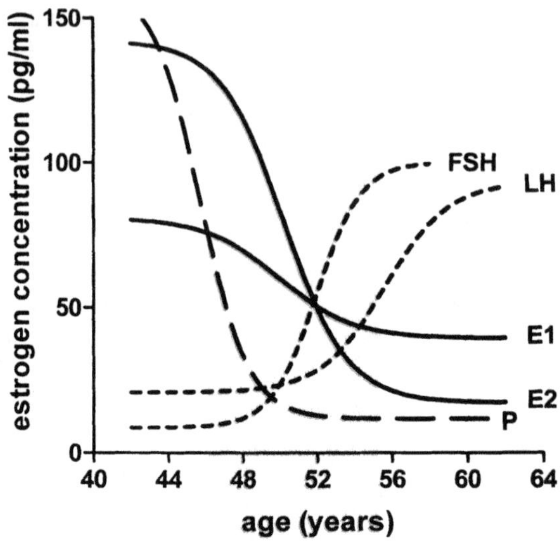

Fig. 5. Schematic drawing of the mean serum levels of E2 and E1 in relation to progesterone (P), LH, and FSH during pre- and postmenopause (modified according to ref. [56])

However, even after menopause there is endogenous estrogen production. One source of small amounts of E2 is the adrenal cortex. In addition, androgen precursors like DHEA, DHEAS, testosterone and androstenedione are the substrates for E1 synthesis by the enzyme aromatase in adipose tissue [48]. There is a positive relationship between the amount of adipose tissue and the aromatase activity. As a consequence, overweight postmenopausal women show higher serum E1 levels than their lean peers. Therefore, some of these women have clear signs of estrogen effects in their target organs.

6.2 Consequences of Decreasing Estrogen Concentrations During Menopause

The earliest symptoms of the imminent menopause are menopausal hot flashes. These are characterized by an abrupt increase of skin temperature associated with tachycardia, flushing of the skin and sweating. They are caused by an abrupt change of the setpoint of the hypothalamic temperature center to a lower level, resulting in peripheral vasodilation and increased sweating that is mediated by sympathetic nerve activity as part of a compensatory mechanism regulating body temperature. It has been demonstrated that the administration of exogenous estrogens but not progestins may reverse the menopausal hot flashes [49].

Apart from these immediate consequences of E2 withdrawal, the target organs suffer from a lack of estrogens and show a number of adverse reactions during the postmenopausal phase. Two major organ systems that are affected by these changes are the cardiovascular system and the skeletal system [50]. The cardiovascular system of the fertile woman is considerably less susceptible to arteriosclerosis than that of their male peers. This is at least partially due to an interaction of estrogens with the nitric oxide system that protects against the emergence of arteriosclerotic plaques [51]. Moreover, there are favorable impacts on plasma lipids as well as anti-platelet and antioxidant effects [52]. As soon as estrogen serum levels drop in the postmenopausal phase, there is a rapid increase in the formation of arteriosclerosis in all arterial vessels and a substantial increase in the risk of myocardial infarction or cerebrovascular thrombosis.

In fertile women between age 30 and 40 years, the annual loss of the bone mass ranges from 0.5 to 1 percent. Due to the lack of E2, this process is accelerated up to 10% in the first year after menopause, if estrogens are not replaced. Apart from the mineral components of the skeletal system, the organic substances, i.e., predominantly collagen type I, are removed [53]. As a consequence, the risk of bone fractures increases dramatically within the first 10–20 years after menopause. Certainly, the loss of bone mass is not exclusively caused by the lack of estrogens. It is also influenced by a lack of physical activity, poor calcium intake and decreasing concentrations of growth hormone.

One novel aspect of the adverse effect of E2 withdrawal during menopause is the impairment of cognitive functions. Basic neuroscience studies have elucidated mechanisms of action of estrogens on the structure and function of brain areas known to be critically involved in memory. Controlled clinical studies show that the administration of estrogens to postmenopausal women enhances verbal memory and maintains the ability to learn new materials. These observations are supported by investigations of healthy, elderly women and by a study in which younger women received a gonadotropin-releasing-hormone analogue that suppressed ovarian function [8, 9].

As a consequence of the collagen type I loss, the skin is affected by the cessation of adequate estrogen supply as well. Water content decreases and the epidermis becomes thinner. There is also an atrophy of the secretory components of the skin. Vagina, uterus and breasts involute as a consequence of estrogen withdrawal. These organs decrease in size and especially the vagina is more susceptible to trauma. Similar changes are seen in bladder and urethra. Due to the involution of these organs, there is a higher incidence of functional urinary incontinence in postmenopausal women.

Because of all these complications, estrogen replacement therapy is generally recommended. The correct hormonal treatment, including careful monitoring of its side effects (like an increased susceptibility for breast cancer), may reduce general mortality by 30–50% [45].

7
Tables of Concentrations of Estrogens During Different Phases of Life

It has to be emphasized that the determination of plasma estrogen levels varies considerably with the method used. Therefore, three conditions are compulsory for any specific assay measuring the three major endogenous estrogens E2, E1 and E3:

(1) reference values must be provided for every assay,
(2) the reference values must not be related to the age but to the different developmental stage, i. e., puberty and menopause, and
(3) the assay must be specific.

The reference values shown below in Tables 1–3 were established using radioimmunoassay after chromatographic separation. Lately, ultrasensitive assays for the determination of E2 concentrations have been introduced. Under certain pathophysiological situations like the premature telarche these new assays allow for discrimination even in prepubertal girls [14].

Table 1. Serum estradiol (E2) concentrations during infancy, childhood, different stages of puberty according to Tanner [54], and adulthood [55]

	Age/Phase	Reference values (conventional units)	Reference values (SI units)
Girls	1 week–7 months	< 7–55 pg/mL	< 26–201 pmol/L
	6–12 months	< 7–44 pg/mL	< 26–162 pmol/L
	2nd year	< 7–24 pg/mL	< 26–88 pmol/L
	2–7 years	< 7–12 pg/mL	< 26–44 pmol/L
	P I and > 7 years	< 7–20 pg/mL	< 26–73 pmol/L
	P II	< 7–35 pg/mL	< 26–129 pmol/L
	P III	7–60 pg/mL	26–220 pmol/L
	P IV	12–93 pg/mL	44–341 pmol/L
	P V	12–250 pg/mL	44–918 pmol/L
Boys	1 week–7 months	< 7–25 pg/mL	< 26–92 pmol/L
	6–12 months	< 7–19 pg/mL	< 26–70 pmol/L
	2nd year	< 7–17 pg/mL	< 26–61 pmol/L
	2–7 years	< 7–11 pg/mL	< 26–40 pmol/L
	P I and > 7 years	< 7–14 pg/mL	< 26–51 pmol/L
	P II	< 7–15 pg/mL	< 26–55 pmol/L
	P III	8–26 pg/mL	29–96 pmol/L
	P IV	9–32 pg/mL	33–118 pmol/L
	P V	12–39 pg/mL	44–142 pmol/L
Women	Follicular phase	30–300 pg/mL	110–1100 pmol/L
	Ovulation	300–400 pg/mL	1100–1450 pmol/L
	Luteal phase	> 130 pg/mL	> 470 pmol/L
	Postmenopause	< 20 pg/mL	< 70 pmol/L
Men		< 50 pg/mL	< 180 pmol/L

Table 2. Serum estrone (E1) concentrations during infancy, childhood, different stages of puberty according to Tanner [54], and adulthood [55]

	Age/Phase	Reference values (conventional units)	Reference values (SI units)
Girls	1 week-7 months	< 7-27 pg/mL	< 26-100 pmol/L
	6-12 months	< 7-16 pg/mL	< 26-60 pmol/L
	2nd year	< 7-14 pg/mL	< 26-52 pmol/L
	2-7 years	< 7-17 pg/mL	< 26-63 pmol/L
	P I and >7 years	< 7-29 pg/mL	< 26-107 pmol/L
	P II	< 7-37 pg/mL	< 26-137 pmol/L
	P III	8-53 pg/mL	26-196 pmol/L
	P IV	10-53 pg/mL	44-285 pmol/L
	P V	12-142 pg/mL	44-525 pmol/L
Boys	1 week-7 months	< 7-21 pg/mL	< 26-78 pmol/L
	6-12 months	< 7-21 pg/mL	< 26-78 pmol/L
	2nd year	< 7-18 pg/mL	< 26-68 pmol/L
	2-X7 years	< 7-13 pg/mL	< 26-48 pmol/L
	P I and >7 years	< 7-19 pg/mL	< 26-70 pmol/L
	P II	11-30 pg/mL	41-111 pmol/L
	P III	11-31 pg/mL	41-115 pmol/L
	P IV	15-41 pg/mL	56-152 pmol/L
	P V	21-47 pg/mL	78-174 pmol/L
Women	fertile phase	20-182 pg/mL	74-673 pmol/L
	postmenopause	15-80 pg/mL	53-280 pmol/L
Men		22-48 pg/mL	81-178 pmol/L

Table 3. Estriol (E3) concentrations in the maternal serum during the course of gestation [55]

Week of gestation	Reference values (conventional units)	Reference values (SI units)
20	1.3-3.2 ng/mL	4.5-11.1 nmol/L
21	1.3-3.6 ng/mL	4.5-12.5 nmol/L
22	1.4-4.0 ng/mL	4.9-13.9 nmol/L
23	1.4-4.4 ng/mL	4.9-15.3 nmol/L
24	1.5-5.0 ng/mL	5.2-17.4 nmol/L
25	1.6-5.2 ng/mL	5.6-18.0 nmol/L
26	1.8-5.6 ng/mL	6.2-19.4 nmol/L
27	2.0-6.0 ng/mL	6.9-20.8 nmol/L
28	2.2-6.5 ng/mL	7.6-22.6 nmol/L
29	2.4-6.9 ng/mL	8.3-23.9 nmol/L
30	2.6-7.2 ng/mL	9.0-25.0 nmol/L
31	2.8-7.7 ng/mL	9.7-26.7 nmol/L
32	2.9-8.4 ng/mL	10.1-29.1 nmol/L
33	3.0-10.0 ng/mL	10.4-34.7 nmol/L
34	3.2-12.0 ng/mL	11.1-41.6 nmol/L
35	3.5-13.8 ng/mL	12.1-47.9 nmol/L
36	4.0-16.0 ng/mL	13.9-55.5 nmol/L
37	4.8-18.0 ng/mL	16.7-62.5 nmol/L
38	5.5-19.5 ng/mL	19.1-67.7 nmol/L
39	6.0-20.0 ng/mL	20.8-69.4 nmol/L
40	6.4-20.3 ng/mL	22.2-70.4 nmol/L
41	6.7-20.0 ng/mL	23.2-69.4 nmol/L
42	6.0-19.5 ng/mL	20.8-67.7 nmol/L

8 References

1. Casey ML, MacDonald PC (1992) Alterations in steroid production by the human placenta. In: Pasqualini JR, Scholler R (eds), Hormones and fetal pathophysiology. Marcel Dekker, New York, p 251
2. Grody WW, Schrader WT, O'Malley BW (1985) Endocrin Rev 3:141
3. Lipsett MB (1986) Steroid hormones. In: Yen SSC, Jaffe RB (eds), Reproductive endocrinology. Saunders, Philadelphia, p 140
4. Couse JF, Korach KS (1999) Endocrin Rev 20:358
5. Bhat RA, Harnish DC, Stevis PE, Lyttle CR, Komm BS (1998) J Steroid Biochem Mol Biol 67:233
6. Couse JF, Lindzey J, Grandien K, Gustafsson JA, Korach KS (1997) Endocrinology 138:4613
7. Taylor JA, Lewis KJ, Lubahn DB (1998) Mol Cell Endocrinol 145:61
8. Sherwin *B* (1998) Proc Soc Exp Biol Med 217:17
9. McEwen BS, Alves SE (1999) Endocrin Rev 20:279
10. Korach KS, Couse JF, Curtis SW, Washburn TF, Lindzey J, Kimbro KS, Eddy EM, Migliaccio S, Snedeker SM, Lubahn DB, Schomberg DW, Smith EP (1996) Recent Prog Horm Res 51:159
11. MacGillivray MH, Morishima A, Conte F, Grumbach M, Smith EP (1998) Horm Res 49 (Suppl 1):2
12. Rey-Stocker I (1997) Weibliche Keimdrüsen. In: Stolecke H (ed), Endokrinologie des Kindes und Jugendalters. Springer, Berlin Heidelberg New York, p 153
13. Kelch RP, Khoury SA, Hale PM (1987) The episodic secretion of hormones. In: Crowley WF, Hoffer JG (eds), Churchill-Livingston, New York, p 187
14. Klein KO, Mericq V, Brown-Dawson JM, Larmore KA, Cabezas P, Cortinez A (1999) J Pediatr 134:190
15. Kulin HE, Grumbach MM, Kaplan SL (1969) Science 166:1012
16. Grumbach MM, Kaplan SL (1990) The neuoendocrinology of human puberty: an ontgenetic perspective. In: Grumbach MM, Sizonenko, Aubert ML (eds), Control of the onset of puberty. Williams and Wilkins, Baltimore, p 1
17. Ruf KB (1973) J Neurol 204:95
18. O'Byrne KT, Chen MD, Nishihara M, Williams CL, Thalabard JC, Hotchkiss J, Knobil E (1993) Neuroendocrinology 57:588
19. Tanner JM (1973) Nature 243:95
20. Tanner JM, Eveleth PB (1985) Changes at age in menarche in Scandinavian countries, 1840–1978. In: Berenberg SR (ed), Puberty, biologic and psychosocial components. HE Stenfert Krose, Leiden, p 256
21. Beyer, Wozniak A, Hutchinson JB (1993) Neuroendocrinology 58:673
22. Balthazart J, Ball GF (1995) Trends Endocrinol Metab 6:21
23. Balthazart J, Foidart A (1983) J Steroid Biochem Mol Biol 44:521
24. Karsch FJ, Dierschke DJ, Knobil E (1973) Science 179:484
25. Lopez FJ, Merchenthaler I, Liposits Z, Negro-Vilar A (1996) Cell Mol Neurobiol 16:129
26. Moore DC, Schlaepfer LV, Paunier L, Sizonenko PC (1984) J Clin Endocrinol Metab 58:492
27. Large DM, Anderson DC (1979) Clin Endocrinol 11:505
28. Large DM, Anderson DC, Laing I (1980) Clin Endocrinol 12:293
29. Berkovitz GD, Guerami A, Brown TR, MacDonald PC, Migeon CJ (1985) J Clin Invest 75:1763
30. Taylor AE, Whitney H, Hall JE; Martin K, Crowley WF Jr (1995) J Clin Endocrinol Metab 80:1541
31. Shoham Z, Schachter M (1996) Fertil Steril 65:687
32. Wira CR, Stern JE (1992) Endocrine regulation of the mucosal immune system in the female reproductive tract. In: Pasqualini JR, Scholler R (eds), Hormones and fetal pathophysiology. Marcel Dekker, New York, p 343

33. Murphy LC, Dotzaw H, Leygue E, Douglas D, Coutts A, Watson PH (1997) J Steroid Biochem Mol Biol 62:363
34. Kuller LH (1995) Public Health Rev 23:157
35. Lenton EA (1988) Pituitary and ovarian hormones in implantation and early pregnancy. In: Chapman M, Grudzinskas G, Chard T (eds), Implantation. Springer, Heidelberg, p 17
36. Dey SK, Paria BC, Andrews GK (1991) Uterine EGF ligand receptor circuitry and its role in embryo-uterine interactions during implantation in the mouse. In: Strauss JF, Lyttle CR (eds), Uterine and embryogenic factors in early pregnancy. Alan R. Liss, New York, p 211
37. Yoshinaga K (1994) Endocrinology of implantation. In: Tulchinsky D, Little AB (eds), Maternal/fetal endocrinology. WB Saunders Company, Philadelphia, p 336
38. Siiteri PK, MacDonald PC (1963) J Clin Endocrinol Metab 26:751
39. Csapo AI, Pulkkinen MO, Wiest WG (1973) Am J Obstet Gynecol 15:759
40. MacDonald PC, Siiteri PK (1965) J Clin Invest 44:465
41. Tulchinsky D, Hobel CJ, Korenman SG (1971) Am J Obstet Gynecol 111:311
42. Lindberg BS, Johannon EDB, Nilsson BA (1974) Acta Obstet Gynecol Scand 32:21
43. Topper YJ (1970) Recent Progr Horm Res 26:287
44. Kjer JJ, Hess J (1983) Acta Obstet Gynecol Scand 62:307
45. Greendale GA, Lee NP, Arriola ER (1999) Lancet 353:571
46. Treolar AE, Boynton RE, Behn BG, Brown DW (1967) Int J Fertil 12:77
47. Judd HL, Koreman SG (1982) Effects of aging on reproductive function in women. In: Koreman SG (ed), Endocrine aspects of aging. Elsevier Biomedical, Amsterdam, p 163
48. Longcope C, Pratt JH, Schneider SH, Fineberg SE (1978) J Clin Endocrinol Metabol 46:146
49. Meldrum DR, Shamonk IM, Frumar AM, Tataryn IV, Shang RJ, Judd HL (1979) Am J Obstet Gynecol 135:713
50. Leidenberger (1992) Endokrinologie der perimenopausalen Übergangsphase, der Menopause und des Seniums. In: Klinische Endokrinologie für Frauenärzte. Springer, Berlin Heidelberg New York, p 101
51. Bell DM, Johns TE, Lopez LM (1998) Ann Pharmacother 32:459
52. Maxwell SR (1998) Basic Res Cardiol 93 Suppl 2:70
53. Brincat M, Monitz CF, Kabalan S (1987) Br J Obstet Gynaecol 94:126
54. Bidlingmaier F, Butenandt O, Knorr D (1977) Pediatr Res 11:91
55. Allolio B, Schulte HM (1996) Praktische Endokrinologie. Urban und Schwarzenberg, München, p 734
56. Rudolf K (1996) Klimakterium-hormonelle Substitution. In: Allolio B, Schulte HM (eds), Praktische Endokrinologie. Urban und Schwarzenberg, München, p 470

CHAPTER 6

Dietary Estrogens of Plant and Fungal Origin: Occurrence and Exposure

W. E. Ward[1], L. U. Thompson[2]

Department of Nutritional Sciences, Faculty of Medicine, University of Toronto, 150 College Street, Toronto, Ontario, Canada, M5S 3E2
[1] e-mail: Wendy.ward@utoronto.ca
[2] e-mail: mailto:lilian.thompson@utoronto.ca

Phytoestrogens and mycoestrogens are naturally occurring dietary compounds that strongly resemble the structure of the mammalian steroidal estrogens. Lignans, isoflavones, and coumestans are the three major classes of phytoestrogens to which humans and animals are exposed. Animals may be exposed to high levels of phytoestrogens while grazing in pastures or consuming feed rich in clover or alfalfa. Domestic livestock may be exposed to mycoestrogens, primarily zearalenone, by consuming feed that is contaminated with *Fusarium* spp. toxins. Livestock consuming feed is exposed to 907–1195 mg of isoflavones/kg of feed. In clover pastures, livestock is exposed to isoflavones at a level of 0.05–4.8% (dry weight). Mycoestrogen exposure varies according to the level of contamination of feed. The zearalenone content of animal feed is estimated to be within 14–215 ng/g depending on the geographical region and type of grain or cereal consumed. With respect to human exposure, lignans and isoflavones are most commonly found in foods containing flaxseed or soybeans, respectively. Since the extent of phytoestrogen exposure is dependent on dietary composition, vegetarians or infants receiving soy-based infant formulas have a significantly higher level of exposure to phytoestrogens. Humans may be exposed to trace or low levels of mycoestrogens via consumption of cereal or cereal products that are mildly contaminated with zearalenone. Dietary intakes of zearalenone are estimated to be 100–500 ng/kg body weight per day. Both animal and human studies have demonstrated that exposure to phytoestrogens and mycoestrogens can result in estrogen-like or antiestrogen-like effects depending on the timing of exposure in the life-cycle, the duration of exposure, and the dose administered. Accordingly, these compounds can have adverse effects or health benefits. Alterations in reproductive indices that lead to reduced fertility rates have been reported in animals grazing in pastures containing phytoestrogens or consuming feed contaminated with mycoestrogens. Impairments in sexual behavior and alterations in measures of masculinity as well as modifications in carcass composition have been reported in animals implanted with a synthetic analogue of zearalenone. In humans, the progression of diseases in which estrogen may play a role, such as cancer, cardiovascular disease, and osteoporosis, may be attenuated with phytoestrogen exposure. Exposure to phytoestrogens during critical developmental periods may reduce the risk of disease development in later life.

Keywords. Phytoestrogens, Mycoestrogens, Lignans, Isoflavones, Disease

1 **Introduction** 102

2 **Occurrence and Metabolism of Dietary Estrogens** 103

2.1 Phytoestrogens 103
2.2 Mycoestrogens: Zearalenone 107

The Handbook of Environmental Chemistry Vol. 3, Part L
Endocrine Disruptors, Part I
(ed. by M. Metzler)

3 **Domestic Animals** 108

3.1 Exposure to Dietary Estrogens 108
3.2 Implications for Domestic Animal Health 109

4 **Humans** 111

4.1 Exposure to Dietary Estrogens 111
4.2 Implications for Human Health 114
4.2.1 Stages of the Life-Cycle 114
4.2.2 Cancer 118
4.2.3 Cardiovascular Disease 120
4.2.4 Osteoporosis 121

5 **Conclusions** 124

6 **References** 125

List of Abbreviations

ED	Enterodiol
EL	Enterolactone
GnRH	Gonadotrophin releasing hormone
IGF-I	Insulin-like growth factor-I
LDL	Low density lipoprotein
LH	Luteinizing hormone
SDG	Secoisolariciresinol diglycoside

1 Introduction

Scientific investigation into the effects of dietary estrogens on animal and human health has proliferated over the past decade. The interest in phytoestrogens, particularly in relation to animal and human health, is due to the estrogen-like biological activity of these compounds. Endogenous estrogen is an important modulator of growth, reproduction, cardiovascular health, and bone metabolism. Thus, whether or not consumption of naturally-occurring estrogens may mimic or antagonize the biological action of endogenous estrogen has direct implications for animal and human health. In addition, three of the major diseases afflicting populations in industrialized countries – cancer, cardiovascular disease, and osteoporosis – are diseases in which estrogen can have a role in disease etiology due to anticarcinogenic, antioxidant, and/or antiresorptive properties, respectively. Thus, elucidating the ability of dietary estrogens to prevent, treat, or perhaps perpetuate the disease process is an active area of investigation. This review will provide the reader with an understanding of the natural abundance of estrogens in animal and human foods, the extent of exposure, the metabolic action, and fate of dietary estrogens, and the implications of consumption of dietary estrogens on normal animal and human health as well as in specific human disease states in which estrogen may be implicated in the disease process.

2
Occurrence and Metabolism of Dietary Estrogens

2.1
Phytoestrogens

The three main classes of phytoestrogens are lignans, isoflavones, and coumestans. As indicated by the term phytoestrogens, these compounds are naturally occurring estrogens of plant origin. A common feature of the lignans, isoflavones, and coumestans is their striking structural similarity to 17β-estradiol and the synthetic estrogen, diethylstilbestrol (Fig. 1).

Lignans are abundant in cereal brans, whole cereals, oilseeds, legumes, fruits, vegetables, and seaweeds, but the richest source of lignans is flaxseed. In vitro fermentation of 68 common plant foods with human fecal microbiota for 48 h

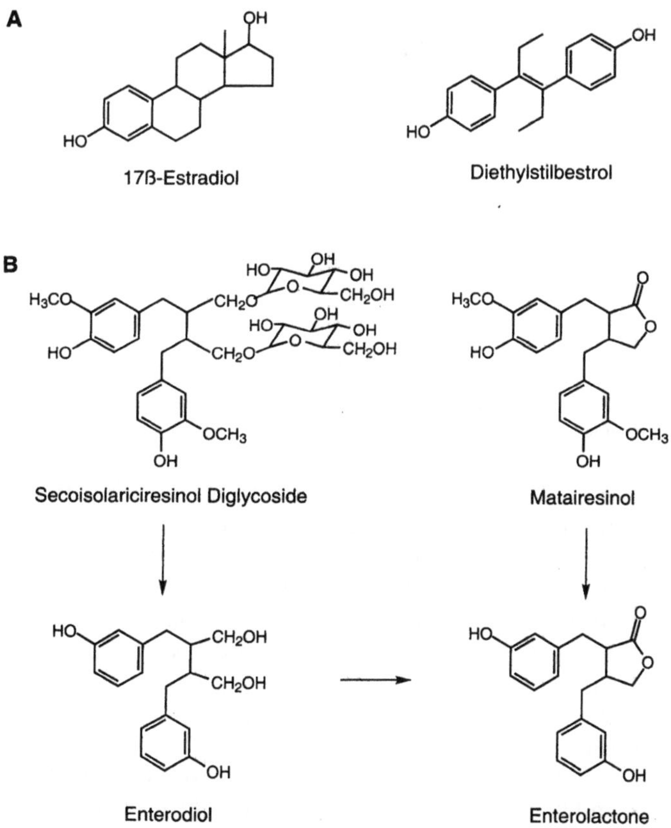

Fig. 1 A, B. A The structure of 17β-estradiol and of the synthetic estrogen, diethylstilbestrol. B Metabolism of secoisolariciresinol diglycoside and matairesinol to the mammalian lignans.

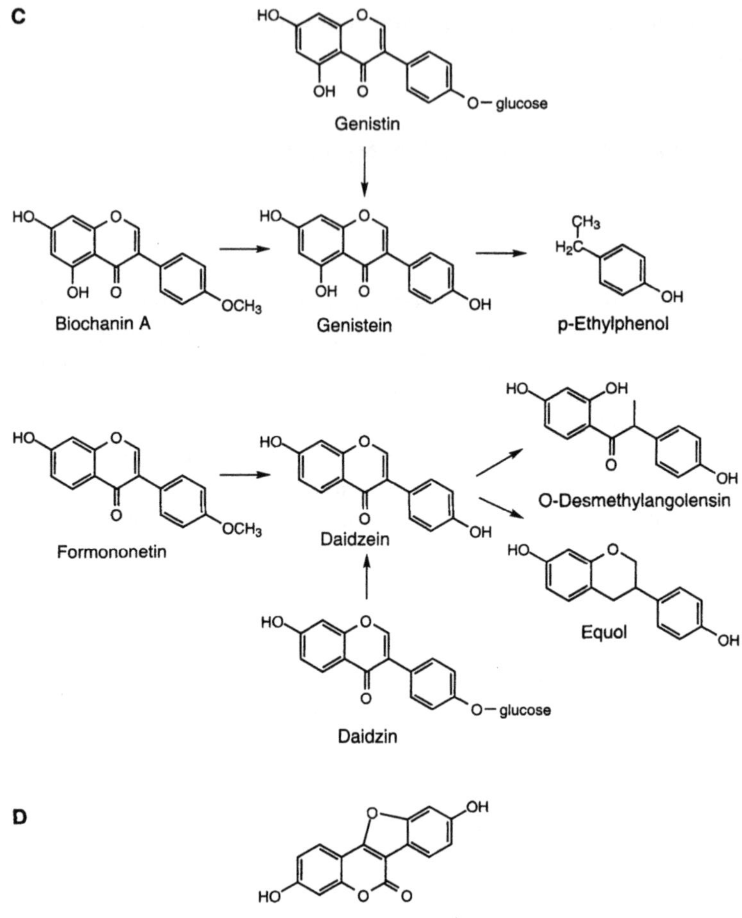

Fig. 1C,D. (continued) **C** Metabolism of the major isoflavones, biochanin A and formononetin. **D** Coumestrol, the major coumestan

demonstrated that flaxseed contains at least 100 times more lignans than other foods, including other oilseeds [1, 2]. Dried seaweed was the second highest lignan-containing food [1]. The lignan concentrations of a variety of plant foods, organized by food group, are summarized in Table 1.

The isoflavone concentrations of a variety of foods have also been measured and reported. While isoflavones are most abundant in soybeans and soy products such as tofu, a variety of beans, sprouts and legumes are also rich sources of isoflavones [2, 3]. Table 2 summarizes a selection of foods that have the highest concentrations of isoflavones.

The major coumestan, coumestrol, is present in several types of sprouts (alfalfa, soy, clover, and bean) as well as bean seeds (pinto, kala chana, and mung) (Table 3). Clover sprouts are a very rich source of coumestrol, containing more than 3 times the level of other foods. Several foods that contain measurable

Table 1. Mammalian lignan production from several plant food groups

Food group	Mammalian Lignans[a,b] (µg/100 g sample[c])
Flaxseed (an oilseed)	67,541
Other Oilseeds	638 ± 219
Dried Seaweeds	900 ± 247
Whole Legumes	562 ± 211
Cereal Brans	486 ± 90
Legume Hulls	371 ± 52
Cereals	359 ± 81
Vegetables	144 ± 23

[a] Data are expressed as the mean of the respective food group ± standard deviation, except for flaxseed.
[b] Lignans include enterolactone and enterodiol as determined using GC/MS.
[c] Sample weight is expressed as a wet weight.
(Summarized from [1]).

Table 2. Plant foods with the highest concentrations of isoflavones

Plant food	Isoflavones[a,b] (mg/kg sample)
Dried soy bean seeds	1953.0
Soy flour	1777.3
Dried black soybean seeds	1310.7
Tofu	278.8
Fresh soy bean seeds	181.7
Dried green split peas	72.6
Clover sprouts	30.6
Dried kala chana seeds	19.0
Soy bean hulls	18.4
Dried black eyed bean seeds	17.3
Dried small white bean seeds	15.6
Dried garbanzo bean seeds	15.2
Dried pink bean seeds	10.5

[a] Data are expressed as the mean value.
[b] Isoflavone concentrations were determined by HPLC and were a mixture of daidzein, genistein, formononetin and biochanin A.
(Modified from [3]).

quantities of coumestrol, such as clover sprouts, pinto bean seeds, and kala chana seeds, are also sources of isoflavones.

Phytoestrogens exist as glycosides in food products. Thus, after ingestion of a phytoestrogen-containing food, bacterial beta-glycosidases in the colon hydrolyze the glycosides into aglycones. Aglycones can be directly absorbed or undergo further metabolism before absorption in the gastrointestinal tract [4]. The bacterial metabolism of these three major classes of phytoestrogens and the structures of the metabolites are illustrated in Fig. 1.

The two predominant lignan precursors are secoisolariciresinol diglycoside (SDG) and matairesinol. SDG and matairesinol are metabolized to the mam-

Table 3. Coumestrol concentrations of various plant foods

Plant food	Coumestrol[a,b] (mg/kg sample)
Clover sprouts	280.6
Dried, round split peas	81.1
Dried kala chana seeds	61.3
Alfalfa sprouts	46.8
Dried pinto beans	36.1
Dried, large, lima bean seeds	14.8
Dried red bean seeds	Trace

[a] Data are expressed as the mean value.
[b] Coumestrol concentrations were determined by HPLC.
(Modified from [3]).

malian lignans enterodiol (ED) [2,3-bis(3-hydroxybenzyl)butane-1,4-diol] and enterolactone (EL) [*trans*-2,3-bis-(3-hydroxybenzyl)butyrolactone], respectively, via a series of dehydroxylation and demethylation reactions. ED, derived from SDG, can be further oxidized to EL by bacteria in the colon.

The major isoflavones are biochanin A (5,7-dihydroxy-4′-methoxy-isoflavone), formononetin (7-hydroxy-4′-methoxy-isoflavone), genistein (4′,5,7-trihydroxy-isoflavone) and daidzein (4′,7-dihydroxy-isoflavone). While genistein and daidzein do exist as inactive glycosides in foods, these compounds can also be formed from biochanin A and formononetin, respectively [5]. Daidzein can also be further metabolized to equol. Coumestrol, the major coumestan, exists in foods as an inactive glycoside and is converted to its aglycone form by colonic bacteria. After absorption, phytoestrogens and/or the corresponding metabolites are conjugated in the liver with glucuronic acid or sulfate, undergo hepatic circulation, and are excreted in the bile [2, 6]. Since a proportion of phytoestrogens and/or their metabolites are excreted in the urine, quantifying urinary lignans is a useful indicator of phytoestrogen intake.

Other plant components reported to have estrogenic activities include the flavonoids (flavones, flavonols, and flavonones), resveratrol, indole-3-carbinol, and unidentified components in various herbs and spices (Fig. 2). In addition, synthetic forms of isoflavones, such as ipriflavone have been reported.

Flavonoids provide pigmentation to seeds, leaves, petals, and fruits of flowering plants. They typically exist as sugar conjugates [7]. Specific flavonoids (4′,7-dihydroxyflavone, 4′,5,7-trihydroxyflavone or apigenin, 4,5,7-trihydroxyflavanone or naringenin) can exert estrogen-like effects by competing with 17β-estradiol for binding to the estrogen receptor [7]. More recently, the effects of resveratrol (3,4′,5-trihydroxystilbene), present in the skin of dark-skinned grapes, and indole-3-carbinol (indole-3-methanol), abundant in cabbage, broccoli, and brussels sprouts, have been studied. Resveratrol exists as a glycoside in grapes and red wine, and its structure is similar to that of diethylstilbestrol, a synthetic estrogen. In humans, it is currently unclear whether resveratrol is absorbed in a sufficient amount to mediate biological effects [8, 9] but administration of moderate levels of red wine containing resveratrol to rats resulted in

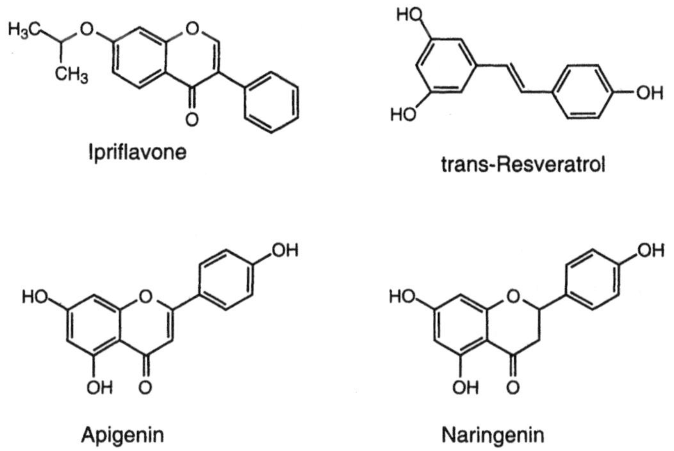

Fig. 2. The structures of the ipriflavone, *trans*-resveratrol, apigenin and naringenin

measurable levels of resveratrol in several different biological tissues including heart, liver, kidney, plasma, and urine [10]. Indole-3-carbinol exists as a glucosinolate and is converted to several biologically active polyaromatic indolic compounds in the acidic environment of the stomach [11]. The ability of 150 different herbs, foods, and spices to bind the estrogen receptor has been recently evaluated [12]. In addition to soy milk and red clover, several herbs and spices, including licorice, mandrake, blood root, thyme, yucca, turmeric, hops, verbenna, yellow dock, and sheep sorrel, were shown to bind the estrogen receptor with binding affinities equivalent to 0.5–2 µg of estradiol per 2 g of dry herb [12].

2.2 Mycoestrogens: Zearalenone

Mycoestrogens are dietary estrogens of fungal origin. Cereals that are stored under moist conditions can be contaminated with molds that produce mycoestrogens [13]. Consequently, both domestic livestock and humans can consume significant quantities of mycoestrogens via the consumption of contaminated rice, corn, wheat, and barley. The major mycoestrogen, zearalenone, is a metabolite of naturally occurring *Fusarium* spp. toxins [14]. Zearalenone is a resorcylic acid lactone [6-(10′-hydroxy-6′-oxo-*trans*-1′-undecenyl)-β-resorcylic acid lactone]. As shown in Fig. 3, zearalenone can be metabolized to zearalanone, zearalenol, and zearalanol.

Several factors have been shown to influence the metabolism of zearalenone to the various metabolites. These factors include the age and sex of the animals, the species, and enzyme activity [14]. Based on in vitro studies, the major enzymes involved in the metabolism of zearalenone are reductase and 3α-hydroxysteroid dehydrogenase enzymes that require NADH or NADPH as coenzymes [14].

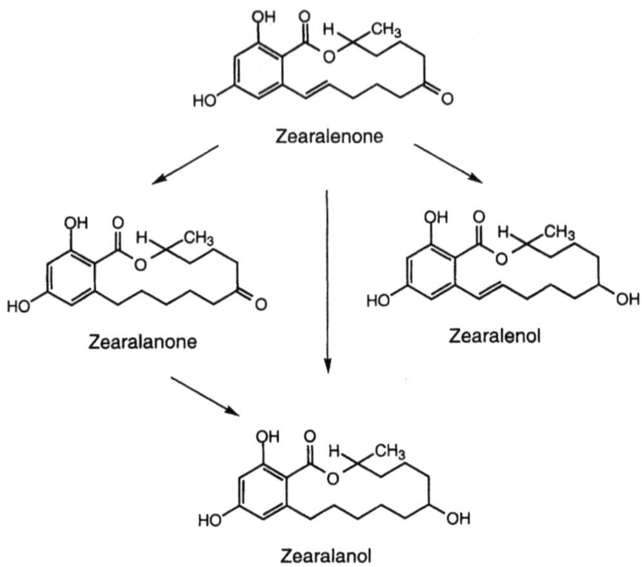

Fig. 3. The structure of zearalenone and its metabolites

3
Domestic Animals

3.1
Exposure to Dietary Estrogens

Domestic livestock can be exposed to both phytoestrogens and mycoestrogens. Animals grazing in pastures can consume many different estrogenic plants such as alfalfa, soybeans, annual medics, as well as white, red, or subterranean clover. Estimates regarding the level of isoflavone exposure to domestic animals has been provided by a Swedish study in which the levels of daidzein, genistein, formononetin, and biochanin A in several major constituents of livestock feed and samples of silage from two different storage sites were quantified [15]. As shown in Table 4, red clover contained more than 10 times the level of isoflavones of the soybean meals.

In comparison to red clover, silage samples contain lower quantities of isoflavones. In pastures containing clover, the level of isoflavones is reported to be 0.05–4.8% (dry weight basis). There are 3 different types of clover: subterranean, red, and white clover with 0.8–4.8%, 0.27–1.95%, and 0.05% isoflavones, respectively. These ranges include the findings from 7 different subterranean, 3 red clover, and 1 white clover genotypes [16–18]. Within a clover plant, the isoflavone concentration is highest in the leaves and lowest in the roots and stems, and intermediate in the flowers [19]. In addition, environmental factors such as temperature, mineral status, water deficiency, or water excess can alter isoflavone concentrations of plants [16, 19–22].

Table 4. Concentration of specific isoflavones in the components of animal feed and two different silage samples

Sample	Daidzein (mg/kg)	Genistein (mg/kg)	Formononetin (mg/kg)	Biochanin (mg/kg)	Total[a] (mg/kg)
Red Clover	530	1060	13220	8330	23140
Toasted and defatted soybean meal	616	753	ND	ND	1369
Whole soybean meal	706	1000	ND	ND	1706
Silage sample A	38	36	562	271	907
Silage sample B	67	52	654	422	1195

[a] Isoflavone concentrations were determined by HPLC.
ND, not detectable.
(Modified from [15]).

There is limited information about the levels of zearalenone or other mycoestrogens in animal feed. Canada's Health Protection Agency reported that the zearalenone content of a variety of grains was 23–215 ng/g [23]. Wheat samples from the Midwestern United States are reported to contain 35–155 ng zearalenone/g [24]. A study from China reports that corn contains 14–169 ng zearalenone/g [25]. The actual level of mycoestrogens that domestic animals are exposed to is likely highly variable, being dependent on the storage and composition of the feed as well as geographical location. Experimental studies, in general, have tested the effects of zearalenone or zeranol, a synthetic derivative of zearalenone, at a concentration of 1 mg/kg body weight [26, 27]. One study in cattle reported that zeranol may be formed in vivo by the presence of the Fusarium spp. toxins. It was reported that Fusarium spp. toxins were present in 32% of bile samples tested [28]. Thus, herd contamination with Fusarium spp. appears to be an additional factor affecting exposure to zearalenone and/or zeranol.

3.2 Implications for Domestic Animal Health

There have been many reports of phytoestrogens disrupting reproductive activity in sheep. One of the earliest reports of the impact of phytoestrogens on domestic animal health arose from the observation that sheep grazing in pastures with a high content of clover experienced fertility abnormalities, including a decreased number of births and prolapse of the uterus [29]. It was later demonstrated that effects on fertility can be temporary or permanent depending on the duration of exposure to phytoestrogens [30, 31]. For example, fertility can be restored if ewes are relocated to a pasture devoid of phytoestrogens if they have been exposed to phytoestrogen-rich pastures for less than 3 years. Temporary infertility is attributed to increased embryo mortality and a reduction or cessation in ovulation. Permanent infertility is purported to occur after 3 years of exposure to dietary estrogenic compounds. This infertility is due to permanent changes in the architecture of the cervix and also changes in the viscoelasticity of the mucus in the cervix which prevents the transport of sperm through the cervix [30, 31]. In addition to these effects on the cervix, phytoestrogens have been shown to exert effects on estrogen-sensitive tissues such as the mammary gland and female reproductive organs of ewes. These estrogen-like effects include the enlargement of the mammary gland and the production of a milky fluid by the mammary glands, hypertrophy of the teats, an increase in the size of the uterus, and an increase in the thickness and keratination of the vaginal epithelium [30, 31]. Isoflavones can also be accumulated in the body fat of animals exposed to phytoestrogen-containing pastures [32]. The impact of this to animal reproduction and the health of humans who consume animal meat are uncertain.

Cattle have also been shown to be sensitive to the estrogen-like effects of dietary phytoestrogens. Specific observations include the swelling of the vulva, discharge of cervical mucus, uterus enlargements, and cystic ovaries. Moreover, irregular estrus cycles including periods of anestrus and decreased rates of conception have been reported [33]. Similar to sheep, these effects are only temporary as the symptoms resolve after placement in a pasture devoid of phytoestrogens [34].

There are two main reasons why the impact of mycoestrogens on animal health have been studied and both are based on the knowledge that mycoestrogens may have weak estrogenic activity. The first reason arises from the fact that animal feed can be contaminated with mycoestrogens that may alter fertility and other reproductive indices. The second reason stems from the economic desire of producing livestock with a more favorable body composition. It is well established that providing anabolic agents to domestic animals (e.g., cattle, swine) effectively influences body composition. Thus, whether the administration of natural, dietary components could also have an anabolic effect has been investigated. The impact of zearalenone consumption on reproductive indices which includes age at puberty onset, pregnancy, and rebreeding, and the number of fetuses or live births per litter, has been extensively studied. In swine, several studies have provided evidence that consumption of zearalenone can prevent pregnancy, stimulate regression of the corpus luteum, impair the development of the blastocyst, and decrease the average number of fetuses per sow [25, 27, 35]. However, interestingly, gilts who were fed zearalenone prior to puberty did not experience any adverse effects. Age at puberty onset, conception and ovulation rates, and the number of fetuses were unaltered compared to gilts not receiving zearalenone [36]. These findings suggest that the timing of exposure may be the critical variable. In addition, piglets born to sows who had received dietary zearalenone throughout pregnancy and lactation did have heavier testes, uterine, and ovarian weights, indicative of an estrogen-like effect. However, the reproductive ability of these offspring was unaffected [37] and provides evidence that zearalenone does not have a sustained effect on offspring exposed to mycoestrogens in utero through lactation.

With respect to carcass composition, lambs implanted with zeranol, a synthetic form of zearalenone, had a more favorable lean to fat tissue ratio at slaughter. The alteration in body composition was attributed to a rise in circulating growth hormone and insulin-like growth factor-I [38]. Growth hormone affects nutrient partitioning, promoting the deposition of lean mass and the mobilization of fat mass. In bulls, zeranol implantation from birth to slaughter also resulted in a greater lean mass compared to untreated bulls [39]. Of potential concern were the findings that behavioral development and masculinity were delayed as evaluated by bunting, mounting attempts, facility rubbing, and a smaller scrotal circumference. There were no measures of reproductive activity in these bulls so it is not known whether the delay in behavioral development or altered masculinity affected reproductive ability.

While agricultural studies have mostly focused on the impact of phytoestrogens and mycoestrogens on domestic animal health, other studies have been conducted in rodents, particularly mice and rats, to gain a more thorough understanding of how phytoestrogens can influence human health. These studies will be discussed in the next section.

4 Humans

4.1 Exposure to Dietary Estrogens

The extent to which a person is exposed to dietary estrogens depends on the type of diet consumed. Thus, exposure may be estimated by calculations based on food intakes or measurement of markers of phytoestrogen intakes such as urinary phytoestrogen excretion or plasma phytoestrogen levels. The isoflavone and lignan content of many foods are known and reported in databases [1, 3, 40]. Therefore, the extent of human exposure to dietary lignans or isoflavones from dietary questionnaires can be calculated and differences in phytoestrogen exposure among populations with diverse dietary habits can be evaluated. Dietary lignan intake and production can be estimated by using the mean in vitro lignan production values from whole grains, fruits, and vegetables [1]. If individuals consume the upper limit of the current recommendations of 8 servings of whole grains and 8 servings of fruits and vegetables, lignan intake is approximately 4.6 μmol (1380 μg) per day [41]. In contrast, the lignan intake and production of vegetarians who have an ideal intake of 9 servings of whole grains, 10 servings of fruit and vegetables, 2 servings of legumes, and 1 serving of oilseeds (e.g., flaxseed) is 7.4 μmol (2220 μg) per day [41]. Thus, it appears that vegetarians have a higher lignan intake and production than non-vegetarians. Interestingly, further calculations demonstrate that, due to the high level of lignans produced from oilseeds, particularly flaxseed, omnivores consuming an additional 2 slices of flaxseed bread with approximately 3 g of flaxseed per slice could raise their intake by 7.4 μmol (2220 μg) of lignans per day and thereby have a higher lignan intake and production than vegetarians. Conversely, diet supplemented with 25 g of flaxseed can increase the lignan intake by 22 mg.

The other way to estimate human exposure to dietary estrogens is to evaluate plasma and/or urinary phytoestrogen concentrations among different populations since these measures are indicators of phytoestrogen exposure. It is reported that 7–30% of ingested isoflavones can be recovered in urine, making it possible to approximate isoflavone intake [42]. As shown in Table 5, both

Table 5. Urinary lignan and isoflavone excretion of premenopausal and postmenopausal women consuming vegetarian, lacto-ovovegeratarian and omnivorous diets

Population	Lignans (μmol/day)	Isoflavones (nmol/day)
Premenopausal Finnish women		
Omnivores	2.89	391
Vegetarians	4.16	665
Postmenopausal Finnish women		
Omnivores	1.99	95.3
Vegetarians	8.09	323
Premenopausal American women		
Omnivores	2.22	515
Lacto-ovovegetarians	6.78	1862
Postmenopausal American women		
Omnivores	2.07	178
Lacto-ovovegetarians	1.09	1282

(Modified from [43]).

Finnish premenopausal and postmenopausal women who followed a vegetarian diet had a higher excretion of lignans and isoflavones, indicating a higher phytoestrogen intake, than women consuming an omnivorous diet [43]. From these urinary excretion data, isoflavone intakes of premenopausal Finnish women are estimated at 2216–9500 nmol per day for vegetarians and 1303–5585 nmol per day for omnivores while postmenopausal vegetarians and omnivores consume 1077–4614 nmol per day and 318–1361 nmol per day, respectively. Similarly, both American premenopausal and postmenopausal lacto-ovovegetarian women had higher isoflavone excretion while urinary lignans were not different compared to omnivores [43]. The calculated isoflavone intakes for these premenopausal American women, based on urinary excretion data, are 1717–7357 nmol per day (omnivores) and 6206–26,600 nmol per day (lacto-ovovegetarians). For postmenopausal American women, the isoflavone intakes are estimated to be 4273–18314 nmol per day (lacto-ovovegetarians) and 593–2542 nmol per day (omnivores).

Other studies have shown that geography and cultural practices can also influence exposure to dietary estrogens. Estimates of the total isoflavone intake or intakes of a specific isoflavone (i.e., genistein) in Japanese populations based on the isoflavone concentration of soy products consumed are variable, being 20–45 mg of isoflavones [42, 44] or 20–80 mg of genistein per day [45, 46]. In Western countries, the isoflavone intake is less than 5 mg per day [44] while the intake of genistein in the United States is estimated at 1–3 mg per day [45, 46]. Japanese women living in Japan were shown to have 1.28 and 2.3 times the level of urinary isoflavones compared to premenopausal and postmenopausal women, respectively, who immigrated from Japan to Hawaii [43, 47]. Isoflavone intakes are estimated to be 1576–6757 nmol per day among Japanese women living in Japan as opposed to 1223–5242 nmol per day in premenopausal and

703–3014 nmol per day in postmenopausal immigrants. The higher urinary phytoestrogen excretion in Japanese women living in Japan may be due to the fact that soybeans and soy-based foods such as tofu, are common and thus very accessible for consumption in Japan compared to other countries. Further, higher rates of hormone-specific cancers, such as breast cancer, in Western compared to Asian countries are partially attributed to lower phytoestrogen intakes. In general, breast cancer patients tend to have lower phytoestrogen intakes, as indicated by lower phytoestrogen excretion, compared to vegetarian or lacto-ovovegetarian women [43]. Premenopausal and postmenopausal women from Finland with breast cancer had urinary lignans that were 1.8- and 3.9-fold lower than vegetarian women [43]. In addition, urinary isoflavone excretion was 2.4- and 3.4-fold lower among Finnish premenopausal and postmenopausal breast cancer patients compared to vegetarians. The isoflavone intakes of these breast cancer patients are estimated to be 930–3986 nmol per day for premenopausal women and 314–1345 nmol per day for postmenopausal women. Thus, the extent of phytoestrogen exposure may have important consequences with respect to disease prevention.

There are limited data pertaining to plasma phytoestrogen levels in maternal and cord blood. In a group of seven Japanese women, isoflavonoid concentrations in maternal and cord blood were 19–744 nmol/l and 58–831 nmol/l while lignan concentrations were negligible [48]. In these same subjects, amniotic fluid concentrations of isoflavonoids were 52–779 nmol/l. It is possible that there is some in utero transfer of isoflavones from the mother to the fetus.

The total isoflavone concentration of five commercially available soy-based infant formulas has been recently reported [49]. Powdered formulas contain 1931 ± 175 mg and 2170 ± 90 mg isoflavones per gram soy protein and liquid formulas contain 2275 ± 455 mg, 2284 ± 38 mg, or 2275 ± 328 mg isoflavones per gram soy protein [49]. Thus, the isoflavone intake of a four-month-old infant consuming 800–1000 ml of soy-based infant formula is between 35 mg and 50 mg per day [49]. Not surprisingly, circulating isoflavones are 10 times higher in soy-fed infants compared to infants fed cow's milk-based infant formula [50]. Breast-fed infants of mothers who consume large quantities of soy or who are vegetarian and thereby consume large quantities of phytoestrogens, are exposing their infants to higher levels than other infants but the exposure to phytoestrogens is much lower than the exposure to infants fed soy-based infant formula (isoflavone intake of 22.5–45 mg per day vs 0.005–0.01 mg per day) [49]. Isoflavones are detectable in breast milk and can be transferred to the infants. Lactating women who were challenged with soy foods experienced a dramatic rise in the daidzein and genistein content of their breast milk [49]. It is currently unclear whether or not exposure to high levels of isoflavones, as can be achieved by feeding soy-based formula, has metabolic consequences but the consumption of soy products in infancy is generally advocated to be safe for children [51], and possibly even protective against adult-onset of disease. However, the long-term effect on whole body metabolism remains unknown as there have been no reports on the long-term safety. Ongoing studies are evaluating the safety of administering phytoestrogen-rich formulas and, later in life, foods which are high in phytoestrogens [52].

The extent to which humans are exposed to mycoestrogens is uncertain although it is suggested that the range of tolerable daily intake is 100–500 ng/kg body weight per day [53]. Because zearalenone is predominantly in cereals and cereal products, individuals with high cereal intakes likely have a higher daily intake of zearalenone.

4.2 Implications for Human Health

As previously mentioned, 17β-estradiol is an important modulator of postnatal growth and development, and is also important for maintaining reproductive, cardiovascular, and bone health. In addition, phytoestrogens appear to have a role in the etiology of hormone-dependent cancers. Phytoestrogens may mediate estrogen-like effects by direct interaction with the estrogen receptor. It has recently been established that there are two different estrogen receptors, α and β, which differ in C-terminal ligand-binding domain and the N-terminal trans-activation domain [54]. Evaluation of the estrogenic potency of compounds for both the α and β estrogen receptor has demonstrated that the major phytoestrogens interact with both receptors [54]. There is also evidence that phytoestrogens may act by estrogen-independent mechanisms, e.g., as antioxidants and hydrogen peroxide scavengers [55, 56] or by interfering with eicosanoid [57, 58] and cytokine production and cell signaling [59, 60]. The specific mechanisms and implications of phytoestrogen action during the various stages of the life-cycle and in disease states are described and summarized in the following four sections.

4.2.1 *Stages of the Life-Cycle*

Endogenous steroidal estrogen is an important regulator of normal metabolism throughout the life-cycle, particularly in women. Physiological levels of estrogen change at various stages of the life-cycle. For instance, endogenous estrogen levels are low during early postnatal life, rising at the end of childhood, and reaching the highest levels during adolescence before declining during the aging process. It is hypothesized that when endogenous levels of estrogen are low, exogenous dietary estrogens may exert estrogen-like effects but when endogenous estrogens are high such as occurs during adolescence and adulthood, dietary estrogens act as anti-estrogens [61]. There may be an optimal range in which to maintain estrogen levels in relation to physiological levels, which undergo dramatic changes at various stages of the life-cycle. Thus, appropriately modifying dietary phytoestrogen intakes may lead to health benefits but potential adverse effects must also be monitored and considered.

The effects of phytoestrogens at various stages of the life-cycle have been most extensively studied in rats. Some of these studies are summarized in Table 6.

Anogenital distance, puberty onset, estrous cycling, growth, sex organ weight, and hormonal profile are indicators of estrogen- or anti-estrogen-like

Table 6. Biological effects of phytoestrogen exposure during various stages of the life-cycle in rats

Phytoestrogen	Stage of development	Gender	Phytoestrogen dose, route of administration, duration of exposure	Biological effects	Reference
Genistein	Gestation	Female	5 or 25 mg/day, injected on day 16–20 in utero	5 mg: ↓ AGD, delayed puberty 25 mg: ↓ birthweight	62
Genistein	Gestation	Male	5 or 25 mg/day, injected on day 16–20 in utero	5 mg: ↓ AGD 25 mg: No effects	62
Coumestrol	Neonatal	Female	0.01 mg/day, injected on PND 1–10	↓ GnRH-LH, ↑ basal LH	
Coumestrol	Neonatal	Male	0.01 mg/day, injected on PND 1–10	↑ GnRH-LH, ↑ basal LH	
Coumestrol	Neonatal	Female	0.7–2.0 mg/day, oral, PND 1–21	↓ Growth, lengthened estrous cycle due to ↑ estrus	64
Coumestrol	Neonatal	Male	0.7–2.0 mg/day, oral, PND 1–10,	↓ Male behavior	65
Genistein	Neonatal	Female	0.1 or 1 mg/day, injected on PND	↑ GnRH-LH 1 mg: ↓ GnRH-LH	66
Genistein	Neonatal	Male	0.1 or 1 mg/day, injected on PND 1–10	0.1 mg: ↑ GnRH-LH 1 mg: ↓ GnRH-LH	66
Purified SDG or Flaxseed	Gestation and Lactation	Female	Fed 5 or 10% flaxseed from start of pregnancy through PND 21 Gavaged 1.5 mg SDG during pregnancy	5%: Delayed puberty onset, ↑uterine and ovarian weight at PND 132 10%: Earlier puberty onset, ↓ AGD, lengthened estrous cycle at PND 50 and 132, ↑ uterine weight at PND 132 SDG: Delayed puberty onset, ↑ uterine and ovarian weight at PND 132	67
Purified SDG or Flaxseed	Gestation and Lactation	Male	Fed 5 or 10% flaxseed from start of pregnancy through PND 21 Gavaged 1.5 mg SDG during pregnancy	5%: No effect 10%: ↓ birthweight and weight gain, ↓ AGD, ↑ sex gland, testes, seminal vesicle and prostate weight at PND 132 SDG: No effect	67

Table 6 (continued)

Phytoestrogen	Stage of development	Gender	Phytoestrogen dose, route of administration, duration of exposure	Biological effects	Reference
Flaxseed	Postweaning to Adult	Female	5 or 10%, fed on PND 21–132	5% and 10%: No effect on puberty onset or estrous cycling	68
Flaxseed	Postweaning to Adult	Male	5 or 10%, fed on PND 21–132	5% and 10%: No effect	68
Flaxseed	Gestation to Adult	Female	5 or 10%, fed from start of pregnancy through PND 132	5%: ↑ Uterine and ovarian weight at PND 132 10%: ↑ Plasma estradiol at PND 50 and 132, ↑ uterine weight at PND 132	68
Flaxseed	Gestation to Adult	Male	5 or 10%, fed from start of pregnancy through PND 132	5%: No effect 10%: ↑ Plasma testosterone, ↑ sex gland, testes, seminal vesicle, prostate weight at PND 132	68
Coumestrol	Weanling	Female	0.7–2 mg/day, fed on PND 1–10	Early puberty onset Lengthened estrous cycle due to prolonged diestrus	69
SDG	Adult	Female	0.75–3 mg/day, orally from PND 57–85	Lengthened estrous cycle due to prolonged diestrus	70
Genistein	Adult	Female	1–10 mg/day or 0.1 mg/day, single injection	1–10 mg: ↓ Progesterone-LH 0.1 mg: ↑ progesterone-LH	71

AGD, anogenital distance; GnRH-LH, gonadotropin releasing hormones induced luteinizing hormone surge; LH, luteinizing hormone; PND, postnatal day; SDG, secoisolariciresinol diglycoside.

activity. It is evident that the biological effects due to exposure to genistein, coumestrol, flaxseed, or purified SDG have profound effects on sex hormones and sexual differentation in males and females. Of interest is the finding that genistein exposure in utero and flaxseed exposure during pregnancy and lactation exerts hormone-like effects that are dose-dependent. Low dose genistein (5 mg) but not high dose genistein (25 mg) resulted in shorter anogenital distance in both male and female rats. Similarly, exposure to 5% flaxseed diet during pregnancy and lactation resulted in delayed puberty onset, an anti-estrogenic-like effect, while exposure to 10% flaxseed resulted in an earlier onset of puberty, an estrogenic-like effect but longer estrous cycles due to prolonged diestrus, an anti-estrogenic effect [67]. The implications of an earlier or delayed onset of puberty and subsequent reproductive activity during the life-cycle are unclear. However, it is clear that the timing of exposure is another important factor controlling the outcome of exposure to phytoestrogens. Exposure to phytoestrogens postweaning through to adulthood did not result in hormone-like effects as puberty onset, estrous cycling, plasma estradiol, or testosterone and sex organ weights were unchanged [68]. Together, these studies confirm that pregnancy and lactation are hormone-sensitive periods of the life-cycle in which developmental changes in the central nervous system and programming of the reproductive tract can be permanently altered. Whether a beneficial or adverse effect occurs depends on the dose of the phytoestrogen, the timing of the exposure, and the endogenous estrogen status of the rat.

In clinical trials, phytoestrogen intakes are associated with alterations in the menstrual cycle. In healthy premenopausal women, consumption of 10 g of flaxseed per day for a period of 3 ovulatory cycles lengthened the luteal phase but did not affect the follicular phase and overall cycle length was unaffected [72]. In addition, the ratio of progesterone to estradiol was significantly higher during the luteal phase, and was largely due to the fact that estradiol levels tended to be lower during the luteal phase during flaxseed supplementation. Lower estradiol levels may have been the result of increased estrogen metabolism or inhibition of ovarian aromatase activity [72]. In contrast, isoflavone supplementation (45 mg per day) increased follicular phase length and the length of the menstrual cycle without lengthening the luteal phase [73]. Although the consequences of altering menstrual cycle length or the time spent in specific phases (e.g., luteal vs follicular phase) have not been confirmed, it is speculated that longer menstrual cycles result in a lesser life-time exposure to estrogen. Therefore, these findings may indicate that modification of the levels of isoflavones in the diet may modulate a women's susceptibility to diseases in which a greater life-time exposure to estrogen is associated with an increased risk of disease development (e.g., cancer).

In postmenopausal women, dietary supplementation with flaxseed (25 g per day) or sprouts (10 g dry seed per day) was shown to alter vaginal cytology [74]. Specifically vaginal cell maturation, a measure of estrogen activity, was shown to increase with phytoestrogen supplementation and provides evidence that dietary phytoestrogens can have an estrogen-like effect in women with low endogenous estrogen production. There are isolated reports that relief from postmenopausal symptoms such as hot flushes [75–77] occurs with phytoestrogen supplementation but to date, randomized controlled trials have not been completed. One ongoing study is using a specially designed menopausal quality of life questionnaire to assess the benefits of soybean or flaxseed supplementation (L. Thompson, unpublished).

4.2.2
Cancer

The hypothesis that phytoestrogens, particularly dietary lignans, may have cancer protective effects is supported by the fact that breast cancer patients and omnivores have a lower urinary lignan excretion than vegetarians [78]. Moreover, epidemiological studies have shown that Asian populations, who consume the highest quantities of phytoestrogens and have the highest urinary lignan excretion, have a sixfold lower incidence of cancer than North Americans and Western Europeans [79, 80]. A recent case-control study has also identified an association between phytoestrogen intake, assessed by measuring urinary phytoestrogen excretion, and the risk of breast cancer [81]. Women with the highest intakes of phytoestrogens had the lowest risk of developing breast cancer after adjusting for other risk factors known to affect breast cancer risk.

Several in vivo feeding studies have provided evidence that dietary estrogens can be protective at specific stages of carcinogenesis. In rats, supplementation of flaxseed at the level of 5% reduced epithelial cell proliferation and nuclear aberrations in the mammary gland [82], and also, reduced mammary tumor incidence and size in carcinogen-treated animals [83, 84]. Supplementation with 2.5% or 5% flaxseed also reduced early markers of colon cancer risk (e.g., epithelial cell proliferation and aberrant crypt foci formation) [85, 86]. While flaxseed contains several compounds which may be cancer-protective, the lignan precursor, SDG, was shown to be the mediator of these observed effects. Feeding diets enriched with SDG at the level present in the 5% flaxseed diet had a similar effect as the diet supplemented with 5% flaxseed (84, 86, 87).

The evidence regarding the anticarcinogenic activity of isoflavones or soybeans has been extensively summarized [42, 88–91]. In brief, the majority of studies have shown that the consumption of soybean products suppressed the number of benign or malignant tumors in rats, hamsters, or mice treated with carcinogens [88, 92–97]. Genistein, in particular, has been shown to have anticarcinogenic activity. For instance, genistein inhibits the growth of human prostate cancer cell lines [93] and also inhibits the proliferative growth of human cancer cell lines [98]. In nude mice, pretreatment of MCF-7 or MDA-MB-468 cells with genistein prior to implantation reduced the tumorigenic potential of the cells [97]. With respect to chemoprevention, the timing of phytoestrogen exposure may be an important consideration. Rats which were exposed to genistein during the neonatal or prepubertal stage of development appear to have some protection against chemically-induced mammary tumors [88, 99–102]. Rats exposed to genistein during early life experienced a longer latency period until mammary tumors developed and, furthermore, there was a lower number of tumors compared to rats which had not been exposed to genistein. One mechanism of genistein action has been suggested to be mediated via alterations in transforming growth factor-β signaling [103].

Other in vitro studies have provided information regarding the potential mechanisms underlying the potential anti-carcinogenic effects of phytoestrogens. Some of these mechanisms are summarized in Table 7.

Less commonly studied dietary estrogens such as resveratrol and indole-3-carbinol have also been shown to have cancer-protective effects. Resveratrol suppressed tumor promoter-induced cell transformation, induced cell death, and activated the expression and activity of p53 [114]. Resveratrol has also been shown to have antiproliferative effects by inhibiting the proliferation of human breast epithelial cells in a dose- and time-dependent manner [115]. Interestingly, resveratrol acted independently of the estrogen receptor. Indole-3-carbinol may also protect against carcinogenesis by an estrogen independent pathway. Indole-3-carbinol inhibited the expression of cyclin-dependent kinase 6, a potential regulator of the cell cycle in human breast cancer cells, in a dose- and time-dependent manner [116]. Addition of indole-3-carbinol also suppressed the incorporation of tritiated thymidine in human MCF-7 breast cancer cells, providing evidence of an antiproliferative effect [116]. Further investigation demonstrated that the addition of indole-3-carbinol and tamoxifen as opposed to adding either compound alone more effectively induced growth arrest in MCF-7 breast cancer cells [117]. Indole-3-carbinol is also reported to

Table 7. Some potential anticarcinogenic mechanisms of phytoestrogens

Mechanism	Reference
Free radical scavenging: prevention of DNA and protein damage and lipid peroxidation	56
Inhibition of steroid binding to the sex steroid binding protein	104
Stimulate the growth of MCF-7 cells in the absence of estradiol and inhibit growth of MCF-7 cells in the presence of estradiol	105
Stimulate DNA synthesis in MCF-7 cells at low concentrations but inhibit DNA synthesis in MCF-7 cells at high concentrations	107
Inhibit proliferation of ZR-75-1 breast cancer cell line	107
Induces pS2 expression in MCF-7 cells	108
Inhibit cellular proliferation and angiogenesis in human and bovine endothelial cells	109
Increase sex hormone binding globulin synthesis by Hep G2 liver cells	110
Inhibit 5α-reductase and 17β-hydroxy steroid dehydrogenase in genital skin fibroblasts	111
Inhibit cell growth by modulation of transforming growth factor-β signaling pathways	103
Inhibit aromatase activity in JEG human choriocarcinoma cells	112
Bind to rat nuclear type II binding site	110
Binds to α-fetoprotein to compete with estradiol and estrone	113

(Modified from [4]).

alter favorably the ratio of 2-hydroxyestrone to 16α-hydroxyestrone and thereby reduce the risk of developing cancer [118–120]. Thus, it appears that a variety of dietary estrogens have anticarcinogenic effects.

4.2.3
Cardiovascular Disease

Some of the strongest evidence to support the fact that estrogen is important for maintaining cardiovascular health comes from studies in which postmenopausal women, with high levels of total cholesterol, low density lipoprotein, and/or triglycerides, experience improvements in their blood lipid profile after starting estrogen replacement therapy. Since the blood lipid profile is a modifiable risk factor that can potentially be controlled by estrogen, clinical studies have looked at the effect of dietary phytoestrogens on blood lipids. A meta-analysis of the relationship between soy protein intake and serum lipids concluded that the consumption of soy protein had a positive effect on blood lipid profile [121]. Total cholesterol, low density lipoprotein (LDL), and triglycerides were all shown to be decreased in people consuming soy protein vs animal protein. This meta-analysis included findings from 38 clinical studies. Since the publication of this report, the findings from several new clinical studies have been published and are in agreement with this meta-analysis [122–124].

While elevations in total cholesterol, LDL, and triglycerides are implicated in the cardiovascular disease process, in vitro and animal studies have provided evidence of how phytoestrogens may act by other mechanisms to modify cardiovascular disease progression, and particularly, the formation of an atherosclerotic plaque (Table 8).

The formation of an atherosclerotic plaque is a complex and a multi-stage process, developing over many years [133]. Briefly, the formation of an atherosclerotic lesion is initiated by an injury to the vessel wall and the subsequent activation of the inflammatory response in which eicosanoids and cytokines stimulate platelet activation and adhesion as well as the proliferation and recruitment of smooth muscle cells to the injured site of the vessel. As summarized in Table 8, it is predominantly genistein and, to a lesser extent, resveratrol which have been shown to have potential cardioprotective effects in vitro. Genistein acts by inhibiting tyrosine kinase activity. Since many of the growth factors and cytokines involved with the inflammatory response bind to tyrosine kinase-dependent receptors, genistein has the potential to inhibit platelet activation and aggregation [59]. Resveratrol inhibits both the cyclooxygenase and lipoxygenase pathways, thereby inhibiting the production of the inflammatory eicosanoids, thromboxanes, and leukotrienes [57, 58]. Other isoflavones, in addition to genistein, as well as purified SDG have been shown to act as antioxidants by preventing the oxidation of LDL [126]. This is important as oxidized LDL is thought to be an initiator of damage to vessel walls. Improvements in vascular tone have also been reported. In female macaques with atherosclerosis, administration of soybean improved coronary artery endothelium-derived vasodilation [131, 132]. These in vitro and animal study findings need to be

Table 8. Some mechanisms of phytoestrogen action in relation to cardiovascular disease

Phytoestrogen	Mechanism	Reference
Genistein, Daidzein, Equol, SDG	Antioxidant effect: prevents oxidation of serum LDL	125, 126
Genistein	Decrease thrombin-induced platelet aggregation	127
Genistein	Inhibit platelet activation by collagen, thromboxane A_2 and ADP	128, 129
Genistein	Inhibit tyrosine kinase activity, thereby blocking the activity of growth factors which stimulate platelet activation and aggregation (e.g. platelet derived growth factor), thrombus formation and the proliferation and recruitment of smooth muscle cells which ultimately leads to the formation of atherosclerotic plaques	59
Genistein	Suppress the formation of nitric oxide and contraction of rat aortic rings	130
Isoflavones	Restoration of vascular tone as assessed by normal vasodilation and endothelial-derived vasomotion in monkeys with atherosclerotic lesions	131, 132
Resveratrol	Inhibit cycloxygenase pathway and the production of thromboxane A_2 by platelets	58
Resveratrol	Inhibit lipoxygenase pathway and the production of leukotrienes by neutrophils	57
Resveratrol	Dose-dependent inhibition of ADP and thrombin-induced platelet aggregation	58

confirmed with in vivo studies before the true effect of dietary phytoestrogens on the progression of cardiovascular disease can be evaluated.

4.2.4
Osteoporosis

Endogenous estrogen is a critical regulator of bone mineral homeostasis, particularly during the pubertal growth spurt when rapid growth of long bones occurs [134–137]. A recent study in both males and females showed that sex steroid hormone levels are correlated with bone mineral density and biochemical markers of bone turnover and, furthermore, that the decline in sex hormone levels that accompanies the aging process corresponds with reductions in bone mineral density [138]. The importance of estrogen for bone maintenance in both males and females has been extensively revealed in male and female patients who have genetic abnormalities involving estrogen synthesis (aromatase enzyme) or estrogen receptor deficiency [135, 139]. In both cases, the lack of normal estrogen synthesis or activity resulted in undermineralized long bones and developmental delay of epiphyseal growth. This finding was particularly interesting in the male patient as it provides direct evidence that estrogen is important for the acquisition of bone mass in males as well as females. During later life, estrogen has an equally important role in maintaining bone mass. The loss of estrogen after menopause is accompanied by a loss of bone mass. Since the phytoestrogens can potentially mimic the biological actions of estrogen, these compounds may potentially stimulate the accretion of bone mass until the end of the growth period and be regulators of bone metabolism after long bone growth is completed.

Most studies that examined the potential therapeutic effects of phytoestrogens have been conducted in postmenopausal women (Table 9) or using the rodent osteoporosis model, the ovariectomized rat (Table 10).

Due to the loss of endogenous estrogen production after the menopause, postmenopausal women can experience rapid bone loss, making this population particularly susceptible to developing osteoporosis. As summarized in Table 9, administration of isoflavone or synthetic isoflavones (ipriflavone) appears to have a positive effect on bone metabolism in postmenopausal women by slowing the loss of bone mineral content at the radius, femur, and/or spine after the menopause. While changes in bone mass have not been evaluated in women receiving flaxseed supplementation, serum tartrate-resistant acid phosphatase, an indicator of bone resorption, was reduced [142].

Several studies, using the ovariectomized rat model have shown that phytoestrogen administration prevented significant bone loss in the femur and lumbar spine (Table 10). This positive effect may be mediated by an elevation in transcripts for insulin-like growth factor-I (IGF-I) in the femur after isoflavone treatment. IGF-I is an essential mediator of bone cell metabolism [150] and

Table 9. Biological effects of phytoestrogens on bone metabolism in humans

Phytoestrogen	Population	Phytoestrogen dose, route of administration, duration of exposure	Biological effects	Ref.
Ipriflavone	Postmenopausal women	200 mg, t.i.d, oral for 2 years	↓ Urinary hydroxy proline. Preserved radial bone mineral density compared to controls	140
Ipriflavone	Premenopausal women, medically-induced hypogonadism	600 mg, t.i.d, oral for 6 months	Femur and lumbar bone mineral density was preserved. No changes in biochemical markers of bone formation or resorption	141
Isolated Soy Protein	Postmenopausal women, hypercholesterolemic	1.39 or 2.25 mg isoflavones/g protein	2.25 mg only: ↑ Lumbar spine bone mineral content and bone mineral density	122
Flaxseed	Postmenopausal women	38 g in the form of bread or muffins/day for 6 weeks	↓ Serum tartrate-resistant acid phosphatase	142

Table 10. Biological effects of phytoestrogens on bone metabolism in rats

Phyto-estrogen	Population	Phytoestrogen dose, route of administration, euration of exposure	Biological effects	Ref.
Ipriflavone	Healthy, Adult Male Rats	200 or 400 mg kg^{-1} body weight/d, oral gavage for 1 month	400 mg only: ↑ femur bone mineral density, improved bone biomechanics (impact strength)	143
Isoflavones	Ovariectomized Rats	Fed 2.5 g isoflavones kg^{-1} diet for 6 weeks	No effect on bone mineral density or biochemical markers of bone turnover	144
Zearalanol	Ovariectomized Rats	1 µg, 2 times $week^{-1}$ by intramuscular injection for 6 weeks	Prevented loss whole body, femur and lumbar spine bone mineral density	144
Coumestrol	Ovariectomized Rats	1.5 µmol, 2 times $week^{-1}$ by intramuscular injection for 6 weeks	Prevented loss of femur bone mineral density ↓ urinary pyridinoline	144
Genistein	Ovariectomized, Lactating Rats	Fed 0.5, 1.6 or 5 mg per day for 2 weeks	0.5 mg: ↑ femur ash weight. Greatest preservation of bone mass observed with 0.5 and 1.6 mg dose	145
Isolated Soy Protein	Ovariectomized Rats	Fed 22.7 g soy protein isolate · 100 g^{-1} in the diet for 1 month	Prevented loss of femur bone mineral density as well as vertebral calcium and phosphorus content compared to ovariectomized rats. ↑ serum alkaline phosphatase ↑ serum tartrate-resistant acid phosphatase	146
Isoflavones	Ovariectomized Rats	Fed 109 or 1088 mg isoflavones for 65 days	1088 mg: Prevented loss of femur bone mineral density 109 and 1088 mg: ↑ urinary hydroxyproline, ↑ Insulin-like growth factor-I transcripts in femur	147, 148
Flaxseed	Healthy Adult Female and Male Rats	1.25, 2.5, 5 or 10% whole flaxseed for 56 days	Femur zinc content decreased with increasing flaxseed dose in males and females ↓ Femur calcium, magnesium, phosphorus in males only ↓ Retention of calcium, magnesium, phosphorus in all flaxseed groups	149

t.i.d., three times per day.

thus, an increase in IGF-I transcripts in the femur provides some insight into a potential mechanism of isoflavone action on bone. Whether other growth factors present in bone are affected by isoflavone administration has not been studied but may provide a clearer understanding of the mechanisms of isoflavone action. There is speculation that genistein may act via the transforming growth factor-β signaling pathway [103].

Using an ovariectomized, lactating rat model, genistein preserved bone mass in a dose-dependent manner. Interestingly, the lowest dose of genistein was most effective, suggesting that genistein has a biphasic effect [145]. Administration of zearalanol and coumestrol reduces bone loss at both the femur and lumbar spine after ovariectomy [144]. The effects of flaxseed supplementation, and specifically phytates on bone are less clear as mineral retention in male and female rats was reduced. The fact that femur zinc content decreased proportionately with increasing flaxseed supplemented supports the fact that phytates affected mineral metabolism [149].

Although bone mineral density is assumed to be a measure of overall bone health, the quality of the bone, as measured by assessing various biomechanical properties of bone, is another factor to be considered. In healthy adult male rats, both bone biomechanics and bone mineral density were improved with ipriflavone treatment [143]. In vitro studies in bone culture systems have provided evidence that phytoestrogens, specifically ipriflavone and resveratrol, can affect both bone formation and bone resorption. Ipriflavone stimulated osteoblast differentiation and the production of mRNA for type 1 collagen, osteopontin, and bone sialoprotein by human bone cells [151]. Similarly, resveratrol stimulates proliferation and differentiation of the osteoblast [152]. In mouse calvaria, femur and tibia cells, ipriflavone inhibited the formation of osteoclasts and prevented the activation of mature osteoclasts [153, 154].

Further research is needed to elucidate the impact of phytoestrogen intake on developing bone, the potential gender differences at various stages of the life-cycle, or whether a dose-dependent relationship exists. Answers to these questions will be critical to determine an optimal dose to maximize bone health throughout the life-cycle.

5 Conclusions

Animals and humans can be exposed to levels of dietary estrogens which can have potent effects on health, either harmful or beneficial, depending on the dose of phytoestrogen, class of compound, stage of the life-cycle, and susceptibility of a disease state. Rigorous assessment of potential toxicological effects, not only on reproduction but also on other endocrine systems, should be carefully examined in order to understand fully the potential for phytoestrogens and mycoestrogens to act as endocrine disruptors.

Acknowledgement. Wendy Ward holds a postdoctoral fellowship from the National Institute of Nutrition, Canada.

6 References

1. Thompson LU, Robb P, Serraino M, Cheung F (1991) Nutr Cancer 16:43
2. Rickard SE, Thompson LU (1997) Phytoestrogens and lignans: effects on reproduction and chronic disease. In: F. Shahidi (ed) Antinutrients and phytochemicals in food. ACS Symposium Series 662. American Chemical Society, Washington, DC, p 273
3. Franke AA, Custer LJ, Cerna CM, Narala K (1995) Proc Soc Exp Biol Med 208:18
4. Thompson LU (1998) Experimental studies on lignans and cancer. Baillieres Clin Endocrinol Metab 12:691
5. Setchell KDR, Adlercreutz H (1988) Mammalian lignans and phytoestrogens:recent studies on their formation, metabolism and biological role in health and disease. In: Rowlands IR (ed) Role of the gut flora, toxicity and cancer. Academic Press, London, p 315
6. Barrett J (1996) Environ Health Persp 104:478
7. Miksicek RJ (1993) Mol Pharm 44:37
8. Bertelli A, Bertelli AA, Gozzini A, Giovannini L (1998) Drugs Exp Clin Res 24:133
9. Soleas GJ, Diamandis EP, Goldberg DM (1997) Clin Biochem 30:91
10. Bertelli AA, Giovannini L, Stradi R, Bertelli A, Tillement JP (1996) Int J Tissue React 18:67
11. Broadbent TA, Broadbent HS (1998) Curr Med Chem 5:469
12. Zava DT, Dollbaum CM, Blen M (1998) Proc Soc Exp Biol Med 217:369
13. Abramson D, Mills JT, Marquardt RR, Frohlich AA (1997) Can J Vet Res 61:49
14. Schoental R (1985) Adv Canc Res 45:217
15. Pettersson H, Kiessling K-H (1984) J Assoc Off Anal Chem 67:503
16. Rossiter RC (1970) Aust Vet J 46:1970
17. Wong E (1973) Plant phenolics. In: Butler GW, Bailey RW (eds) Chemistry and biochemistry of herbage, vol 1. Academic Press, London, p 265
18. Cox RI (1978) Plant estrogens affecting livestock in Australia. In: Keeler RF, Van Kampen KR, James LF (eds) Effects of poisonous plants on livestock. Academic Press, New York, p 451
19. Rossiter RC, Beck AB (1967) Aust J Agric Res 18:23
20. Rossiter RC (1973) Aust J Agric Res 23:419
21. Rossiter RC, Barrow NJ (1972) Aust J Agric Res 23:411
22. Rossiter RC, Barrett DW, Klein L (1973) Aust J Agric Res 24:59
23. Scott PM (1997) Food Addit Contam 14:333
24. Hagler WM, Tyczkowska K, Hamilton PB (1984) Appl Environ Microbiol 47:151
25. Luo Y, Yoshizawa T, Katayama T (1990) Appl Environ Microbiol 12:3723
26. Diekman MA, Long GG (1989) Am J Vet Res 50:1224
27. Long GG, Turek J, Diekman MA, Scheidt AB (1992) Vet Pathol 29:60
28. Kennedy DG, Hewitt SA, McEvoy JD, Currie JW, Cannavan A, Blanchflower WJ, Elliiot CT (1998) Food Addit Contam 15:393
29. Bennetts HW, Underwood EJ, Shier FL (1946) Aust Vet J 22:2
30. Adams NR (1995) Proc Natl Acad Sci 208:87
31. Adams NR (1995) J Anim Sci 73:1509
32. Lidner HR (1967) Aust J Agric Res 18:305
33. Adler JH, Trainin D (1960) Refuah Vet 17:115
34. Kallela K, Heinonen K, Saloniemi H (1984) Nord Veterinaermed 36:124
35. Young LG, Ping H, King GJ (1990) J Anim Sci 68:15
36. Rainey MR, Tubbs RC, Bennett LW, Cox NM (1990) J Anim Sci 68:2015
37. Yang HH, Aulerich RJ, Helferich W, Yamini B, Chou KC, Miller ER, Bursian SJ (1995) J Appl Toxicol 15:223
38. Hufstedler GD, Gillman PL, Carstens GE, Greene LW, Turner ND (1996) J Anim Sci 74:2376

39. Unruh JA, Gray DG, Diekman MA (1986) J Anim Sci 62:279
40. Mazur WM, Duke JA, Wahala K, Rasku S, Adlercreutz H (1998) J Nutr Biochem 9:193
41. Nesbitt PD, Thompson LU (1997) Nutr Cancer 29:222
42. Messina M (1995) Am J Clin Nutr 62:645
43. Herman C, Adlercreutz T, Goldin BR, Gorbach SL, Hockerstedt KAV, Watanabe S, Hamalainen EK, Markkanen MH, Makela TH, Wahala KT, Hase TA, Fotsis T (1995) J Nutr 125:757S
44. Coward L, Barnes NC, Setchell KDR, Barnes S (1993) J Agric Food Chem 41:1961
45. Barnes S, Peterson TG, Coward L (1995) J Cell Biochem 22:181
46. Tham DM, Gardner CD, Haskell WL (1998) J Clin Endocrinol Metab 83:2223
47. Goldin BR, Adlercreutz H, Gorbach SL, Woods MN, Dwyer JT, Conlon T, Bohn E, Gershoff SN (1986) Am J Clin Nutr 44:945
48. Adlercreutz H, Yamada T, Wahala K, Watanabe S (1999) Am J Obstet Gynecol 180:737
49. Setchell KDR, Zimmer-Nechemias L, Cai J, Heubi JE (1998) Am J Clin Nutr 68:1453S
50. Setchell KDR, Zimmer-Nechemias L, Cai J, Heubi JE (1997) Lancet 350:23
51. Klein KO (1998) Nutr Rev 56:193
52. Joannou GE, Silink M, McVeagh P, Kelly GE, Waring MA (1998) Am J Clin Nutr 68:1535S
53. Kuiper-Goodman T (1989) Can J Physiol Pharmacol 68:1017
54. Kuiper GGJM, Lemmen JG, Carlsson B, Corton JC, Safe SH, Van Der Saag PT, Van Der Burg B, Gustafsson J-A (1988) Endocrinol 139:4252
55. Yuan YV, Rickard SE, Thompson LU (1999) Nutr Res (in press)
56. Prasad K (1997) Mol Cell Biochem 168:117
57. Goldberg DM, Soleas GJ, Hahn SE, Diamandis EP, Karumanchiri A (1996) Identification and assay of trihydroxystilbenes in wine and their biological properties. In: Watkins T (ed) Wine composition and health benefits. American Chemical Society, Washington D.C.
58. Pace-Asciak CR, Hahn S, Diamandis EP, Soleas G, Goldberg DM (1995) Clin Chim Acta 235:207
59. Akiyama T, Ishida J, Kakagawa S, Ogawara H, Watanabe S, Itoh N, Shibuya M, Fukami Y (1987) J Biol Chem 262:5592
60. Raines EW, Russell R (1995) J Nutr 125:624S
61. Clarkson TB, Anthony MS, Hughes CL (1995) Endocrinol Metab 6:11
62. Levy JR, Faber KA, Ayyash L, Hughes CL Jr (1995) Proc Soc Exp Biol Mol 208:60
63. Register B, Bethel MA, Thompson N, Walmer D, Blohm P, Ayyash L, Hughes CL Jr (1995) Proc Soc Exp Biol Mol 208:72
64. Whitten PL, Lewis C, Naftolin F (1993) Biol Reprod 49:1117
65. Whitten PL, Lewis C, Russel E, Naftolin F (1995) J Nutr 125:771
66. Faber KA, Hughes CL Jr (1991) Biol Reprod 45:649
67. Tou JCL, Chen J, Thompson LU (1998) J Nutr 128:1861
68. Tou JCL, Chen J, Thompson LU (1999) J Toxicol Environ Health 56:555
69. Whitten PL, Naftolin F (1992) Steroids 57:98
70. Orcheson LJ, Rickard SE, Seidl MM, Thompson LU (1998) Cancer Lett 125:69
71. Hughes CL Jr, Chakinala MM, Reece SG, Miller RN, Schomberg DW Jr, Basham KB (1991) Reprod Toxicol 5:127
72. Phipps WR, Martini MC, Lampe JW, Slavin JL, Kurzer MS (1993) J Clin Endocrinol Metab 77:1215
73. Cassidy A, Bingham S, Setchell KDR (1994) Am J Clin Nutr 60:333
74. Wilcox G, Wahlqvist ML, Burger HG, Medley G (1990) Brit Med J 301:905
75. Albertazzi P, Pansini F, Bonaccorsi G, Zanotti L, Forini E, De Aloysio D (1998) Obs Gynecol 91:6
76. Murkies AL, Lombard C, Straus BJG (1995) Maturitas 21:189
77. Adlercreutz H, Hamalainen E, Gorbach S, Goldin B (1992) Lancet 339:1233
78. Adlercreutz H, Fotsis T, Bannwart C, Hamalainen E, Bloigu S, Ollus A (1986) J Steroids Biochem Mol Biol 25:791
79. Henderson BE, Bernstein L (1991) Breast Cancer Res Treat 18:S11

80. Adlercreutz H, Honjo H, Higashi A, Fotsis T, Hamalainen E, Hasegawa T, Okada H (1991) Am J Clin Nutr 54:1093
81. Ingram D, Sanders K, Kolybaba M, Lopez D (1997) Lancet 350:990
82. Serraino M, Thompson LU (1991) Cancer Lett 60:135
83. Serraino M, Thompson LU (1992) Nutr Cancer 17:153
84. Thompson LU, Rickard S, Orcheson L, Seidl M (1996) Carcinogenesis 17:1373
85. Serraino M, Thompson LU (1992) Cancer Lett 63:159
86. Jenab M, Thompson LU (1996) Carcinogenesis 17:1343
87. Thompson LU, Seidl M, Rickard SE, Orcheson L, Fong HHS. (1996) Nutr Cancer 26:159
88. Barnes S (1997) Breast Cancer Res Treat 46:169
89. Kennedy AR (1995) J Nutr 125:733S
90. Adlercreutz H, Mazur W (1997) Ann Med 29:95
91. Messina M, Persky V, Setchell KD, Barnes S (1994) Nutr Cancer 21:113
92. Billings PC, Newberne P, Kennedy AR (1990) Carcinogenesis 11:1083
93. Peterson G, Barnes S (1993) Prostate 22:335
94. Finlay EMH, Wilson DW, Adlercreutz H, Griffiths K (1991) J Endocrinol 129:49
95. St. Clair, Billings PC, Carew J, Keller-McGrandy C, Newberne P, Kennedy AR (1990) Cancer Res 50:580
96. St. Clair WH, Billings PC, Kennedy AR (1990) Cancer Lett 52:145
97. Constantinou AI, Krygier AE, Mehta RR (1998) Am J Clin Nutr 68:1426S
98. Peterson G, Barnes S (1991) Biochem Biophys Res Commun 179:661
99. Fritz WA, Coward L, Wang J, Lamartiniere CA (1998) Carcinogenesis 19:2151
100. Murrill WB, Brown NM, Zhang JX, Manzolillo PA, Barnes S, Lamartiniere CA (1996) Carcinogenesis 17:1451
101. Lamartiniere CA, Murrill WB, Manzolillo PA, Zhang JX, Barnes S, Zhang X, Wei H, Brown NM (1998) Proc Soc Exp Biol Med 217:358
102. Lamartiniere CA, Moore J, Holland M, Barnes S (1995) Proc Soc Exp Biol Med 208:120
103. Kim H, Peterson TG, Barnes S (1998) Am J Clin Nutr 68:1418S
104. Martin ME, Haourigui M, Pelillero C, Benassayag C, Nunez EA (1996) Life Sci 58:429
105. Mousavi Y, Adlercreutz H (1992) J Steroid Biochem Mol Biol 41:615
106. Wang C, Kurzer MS (1997) Nutr Cancer 28:236
107. Hirano T, Fukuoka K, Oka K, Naito T, Hosaka K, Mitsuhashi H, Matsumoto Y (1990) Cancer Invest 8:595
108. Sathyamoorthy N, Wang TTY, Phang TM (1994) Cancer Res 54:957
109. Fotsis T, Pepper MS, Aktas E, Breit S, Rasku S, Adlercreutz H, Wahala K, Montesano R, Schweigerer L (1997) Cancer Res 57:2916
110. Adlercreutz H, Mousavi Y, Clark J (1992) J Steroid Biochem Mol Biol 41:331
111. Evans B, Griffiths K, Morton M (1995) J Endocrinol 147:295
112. Adlercreutz H, Bannwart C, Wahala K (1993) J Steroid Biochem Mol Biol 44:147
113. Garreau B, Vallette G, Adlercreutz H, Wahala K, Makela T, Benassayag C, Nunez EA (1991) Biochim Biophys Acta 1094:339
114. Huang C, Ma WY, Goranson A, Dong Z (1999) Carcinogenesis 20:237
115. Mgbonyebi OP, Russo J, Russo IH (1998) Int J Oncol 12:865
116. Cover CM, Hsieh SJ, Tran SH, Hallden G, Kim GS, Bjeldanes LF, Firestone GL (1998) J Biol Chem 273:3838
117. Cover CM, Hsieh SJ, Cram EJ, Hong C, Riby JE, Bjeldanes LF, Firestone GL (1999) Cancer Res 59:1244
118. Bradlow HL, Michnovicz JJ, Halper M, Miller DG, Wong GY (1994) Cancer Epidemiol Biomarkers Prev 3:591
119. Michnovicz JJ, Adlercreutz H, Bradlow HL (1997) J Natl Cancer Inst 89:718
120. Michnovicz JJ (1998) Int J Obes Relat Metab Disord 22:227
121. Anderson JW, Johnstone BM, Cook-Newell ME (1995) N Engl J Med 333:276
122. Potter SM, Baum JA, Teng H, Stillman RJ, Shay NF, Erdman JW (1998) Am J Clin Nutr 68:1375S
123. Nilausen K, Meinertz H (1998) Am J Clin Nutr 68:1380S

124. Wong WW, O'Brien Smith O, Stuff JE, Hachey DL, Heird WC, Pownell HJ (1998) Am J Clin Nutr 68:1385S
125. Hodgson JM, Croft KD, Puddey IB, Mori TA, Beilin LJ (1996) J Nutr Biochem 7:664
126. Prasad K (1999) Circulation 99:1355
127. Asahi M, Yangi S, Ohta S, Inazu T, Sakai K, Takeuchi F, Taniguchi T, Yamamura H (1992) FEBS Lett 309:10
128. Nakashima S, Koike T, Nozawa Y (1991) Mol Pharmacol 39:475
129. Sargeant P, Farndale RW, Sage SO (1993) J Biol Chem 268:18,151
130. Marczin N, Papapetropoulos A, Catravas JD (1993) Am J Physiol 265:1014
131. Williams JK, Adams MR, Herrington DM, Clarkson TB (1992) J Am Coll Cardiol 20:452
132. Williams JK, Honore EK, Washburn SA, Clarkson TB (1994) J Am Coll Cardiol 24:1757
133. Steinberg D, Parthasarathy S, Carew TE, Witztum JL. (1989) New Engl J Med 320:915
134. Cutler GB (1997) J Steroid Biochem Mol Biol 61:141
135. Frank GR (1995) Acta Pediatr 84:627
136. MacGillivray MH, Morishima A, Conte F, Grumbach M, Smith EP (1998) Horm Res 49:2S
137. Zachmann M, Prader A, Sobel EH, Crigler JF, Ritzen EM, Atares M, Ferrandez A (1996) J Pediatr 108:694
138. Khosla S, Melton LJ, Atkinson EJ, O'Fallon WM, Klee GG, Riggs BL (1998) J Clin Endocrinol Metab 83:2266
139. Simpson ER, Zhao Y, Agarwal VR, Michael MD, Bulun SE, Hinshelwood MM, Graham-Lorence S, Sun T, Fisher CR, Qin K, Mendelson CR (1997) Rec Prog Horm Res 52:185
140. Adami S, Bufalino L, Cervetti R, DiMarco C, DiMunno O, Fantasia L, Isaia GC, Serni U, Vecchiet L, Passieri M (1997) Osteoporosis Int 7:119
141. Gambacciani M, Spinetti A, Piaggesi L, Cappagali B, Taponeco F, Manetti P, Weiss C, Teti GC, La Commare P, Facchini V (1994) Bone Min 26:19
142. Arjmandi BH, Juma S, Lucas EA, Wei LL, Venkatesh S, Khan DA (1998) Effects of flaxseed supplementation on bone metabolism in postmenopausal women. In: Carter JF (ed) Proceedings of the 57th Flax Institute of the United States. North Dakota State University Press, Fargo, North Dakota, p 65
143. Civetelli R, Abbasi-Jarhomi SH, Halstead LR, Dimarogonas A (1995) Calcif Tiss Int 56:215
144. Draper CR, Edel MJ, Dick IM, Randall AG, Martin GB, Prince RL (1997) J Nutr 127:1795
145. Anderson JJB, Ambrose WW, Garner SC (1998) Proc Soc Exp Biol Med 217:345
146. Arjmandi BH, Alekel L, Hollis BW, Amin D, Stacewicz-Sapuntzakis M, Guo P, Kukreja SC (1996) J Nutr 126:161
147. Arjmandi BH, Getlinger MJ, Goyal NV, Alekel L, Hasler CM, Juma S, Drum ML, Hollis BW, Kukreja SC (1998) Am J Clin Nutr 68:1358S
148. Arjmandi BH, Birnbaum R, Goyal NV, Getlinger MJ, Juma S, Alekel L, Hasler CM, Drum ML, Hollis BW, Kukreja SC (1998) Am J Clin Nutr 68:1364S
149. Kaup SM, Hight SC, Ahn SM, Rader JI (1998) Flaxseed and mineral metabolism in rats. In: Carter JF (ed) Proceedings of the 57th Flax Institute of the United States. North Dakota State University Press, Fargo, North Dakota, p 29
150. Rosen CJ, Donahue LR, Hunter SJ (1994) Proc Soc Exp Biol Med 206:83
151. Cheug S-L, Zhang S-F, Nelson TL, Warlow PM, Civitelli R (1994) Calcif Tissue Int 55:356
152. Mizutani K, Ikeda K, Kawai Y, Yamori Y (1998) Biochem Biophys Res Comm 253:859
153. Notoya K, Yoshida K, Shirakawa Y, Taketomi S, Tsuda M (1995) Bone 16:349S
154. Notoya K, Yoshida K, Taketomi, Yamazaki I, Kumegawa M (1995) Calcif Tiss Int 53:206

CHAPTER 7

Alkylphenols and Bisphenol A as Environmental Estrogens

Caroline M. Markey, Cheryl L. Michaelson, Carlos Sonnenschein, Ana M. Soto

Department of Anatomy and Cellular Biology, 136 Harrison Avenue, Tufts University School of Medicine, Boston, MA 02111, USA
e-mail: ana.soto@tufts.edu

A diverse number of chemicals present in the environment may be detrimental to the development and reproduction of wildlife and humans. These chemicals exert their effect through mimicking endogenous estrogens. Two xenoestrogens that are currently produced in large volumes are alkylphenols and bisphenol A (BPA). These chemicals demonstrate estrogenic activity; they increase the proliferation of estrogen target cells, induce estrogen-specific genes and reporter genes, increase the wet weight of the uterus, and induce proliferation of the epithelium in the endometrium and vagina. Alkylphenols are widely distributed through their use as antioxidants and in the synthesis of alkylphenol polyethoxylates for detergents. The release of these chemicals into natural waters and wastewater treatment plants results in exposure of aquatic wildlife. The extent of exposure to non-aquatic organisms is unknown, but it is likely that exposure occurs in species that eat contaminated fish. Humans are exposed primarily through the use of spermicides containing nonoxynol. BPA is used in the packaging of food and beverages, and in health-related products. This chemical and its derivatives leach from such polycarbonate and epoxy resin products leading to exposure of humans predominantly. Evidence from field studies and laboratory experiments indicate that alkylphenols and BPA have the potential to cause ecological problems and affect human health. Degradation products of alkylphenol polyethoxylates have caused feminization of fish in effluent polluted rivers, and can alter reproductive parameters in rodents. BPA is able to induce feminization of neonatal amphibia, proliferative activity in the uterus and mammary glands, alterations in the neuroendocrine axis, and compromise fertility. In utero exposure to this chemical causes alterations in the onset of sexual maturity in females and changes in the development of male reproductive organs. The most disturbing findings reveal that low doses of BPA, which are physiologically relevant to human exposure, cause the most profound biological effects. These data attest to the urgent need for re-evaluating issues of production, use, and waste treatment programs pertaining to all endocrine disrupting chemicals.

Keywords. Alkylphenols, Alkylphenol polyethoxylates, Bisphenol A, Xenoestrogens, Endocrine disruptors

1 **Introduction** 131

2 **Alkylphenols** 132

2.1 Identification of Estrogenic Activity 132
2.2 Production and Uses 133
2.2.1 Alkylphenol Polyethoxylates 133

The Handbook of Environmental Chemistry Vol. 3, Part L
Endocrine Disruptors, Part I
(ed. by M. Metzler)

2.2.1.1 Industrial Uses 133
2.2.1.2 Institutional Uses 134
2.2.1.3 Household Uses 134
2.3 Human Exposure 134
2.4 Environmental Release 134
2.5 Biodegradation 135
2.6 Bioaccumulation and Metabolism 137
2.7 Biological Effects 138
2.8 Developmental and Reproductive Effects 138

3 **Bisphenol A** 139

3.1 History 139
3.2 Identification of Estrogenic Activity 140
3.3 Production and Use 141
3.3.1 Industrial Production Levels 141
3.3.2 Food and Beverage Containers 141
3.3.3 Dental Sealants and Composites 141
3.3.4 Medical Materials 142
3.4 Human Exposure 143
3.5 Environmental Release 143
3.6 Bioaccumulation 143
3.7 Metabolism 144
3.8 Biodegradation 144
3.9 Biological Effects 145
3.9.1 Teratogenic Effects 145
3.9.2 Developmental Effects 145
3.9.3 Reproductive Effects 147
3.9.4 Neuroendocrine Axis 148

4 **Conclusion** 148

5 **References** 150

List of Abbreviations

BADGE	bisphenol A diglycidyl ether
BisGMA	bisphenol A-diglycerol methacrylate
BPA	bisphenol A, 2,2-bis(*p*-hydroxyphenyl)-propane
BPA-DMA	bisphenol A-dimethacrylate
p,p'-DDE	2,2-bis(*p*-chlorophenyl)-1,1-dichloroethene
DDT	2,2-bis(*p*-chlorophenyl)-1,1,1-trichloroethane
o,p'-DDT	2-(*o*-chlorophenyl)-2-(*p*-chlorophenyl)-1,1,1-trichloroethane
DES	diethylstilbestrol
DHT	5α-dihydrotestosterone
ER	estrogen receptor

ERE	estrogen response element
FSH	follicle stimulating hormone
PCBs	polychlorinated biphenyls
TEGDMA	triethyleneglycol dimethacrylate

1 Introduction

Chemicals can act as endocrine disruptors in a variety of ways. They can directly mimic endogenous hormones, antagonize the natural actions of endogenous hormones, or change the rate of synthesis and metabolism of natural hormones. They may also have the capacity to alter hormone receptor levels [1].

There is evidence to suggest that a single chemical can act through more than one pathway to disrupt the normal hormone balance. One chemical that acts as both a partial estrogen agonist and as an androgen antagonist is *p,p'*-DDE (2,2-bis(*p*-chlorophenyl)-1,1-dichloroethene), a metabolite of DDT (2,2-bis(*p*-chlorophenyl)-1,1,1-trichloroethane) [2]. Furthermore, some phthalate esters are estrogen agonists [3, 4] while diethylhexyl phthalate can alter estrogen synthesis in the ovary, resulting in the suppression of estrous cycles [5]. These findings suggest that endocrine disruptors may utilize pathways other than the classical receptor pathway. The focus of this chapter is on chemicals that have been defined as estrogen mimics, yet are suspected to produce endocrine effects that are not mediated by the classical pathway.

Ovarian estrogens are necessary for the development of the female genital tract, the neuroendocrine tissues, and the mammary glands. During adulthood, they exercise primary endocrine control of the ovarian cycle, pregnancy, and nursing. At the cellular level, estrogens regulate the production and secretion of cell-specific proteins and control the proliferation of cells in the female secondary sex organs [6]. These hormones exert their control primarily through two estrogen receptors (ER), ER-α [7] and ER-β [8]; however, ER-α is the most widely distributed receptor in the female genital tract [9, 10]. Evidence suggests that a lifetime of exposure to ovarian estrogens may be a principal risk factor in the development of breast cancer. Similarly, excessive exposure during in utero development results in irreversible alterations in the structure and function of the female genital tract [11, 12].

Estrogens play an important role in the development of the male genital tract as well. This has become evident on the basis of experimental and clinical studies. Mice lacking the ER-α have been shown to exhibit a reduction in testicular weight and sperm counts, resulting in compromised fertility [10]. One clinical study described a case of male infertility in an adult patient who lacked expression of ER-α; his phenotype included an absence of epiphyseal plate fusion and osteoporosis [13]. In rodents, normal variations in exposure to endogenous estrogens during development have been shown to influence adult behavioral patterns [14]. These subtle differences in estrogen levels occur as a result of fetal positioning in utero. Estrogen levels vary depending on whether a fetus is positioned between two male siblings, two female siblings, or one of each sex. Exposure of men to excessive estrogens results in symptoms such as gynecomastia (the development of female-like breasts), decreased libido and impotence, decreased blood androgen levels, and lowered sperm counts [15]. Such effects are assumed to result from interference with the normal function of the hypothalamic-hypophyseal-gonadal axis.

Once natural estrogens were isolated and the synthesis of steroidal and non-steroidal agonists was achieved, clinicians used these compounds for therapeutic purposes. Between the years 1948 and 1971, the synthetic estrogen diethylstilbestrol (DES) was prescribed to millions of women as an anti-abortive measure in spite of a lack of data to justify its use [16]. Decades later it was found that this experiment resulted in severe genital tract malformations and clear cell carcinoma of the vagina in women who were exposed in utero [17]. The use of DES during pregnancy was banned subsequent to this finding. DES is still being used today, albeit less frequently than in the past, for the treatment of breast cancer and to suppress androgen production in prostate cancer patients [18–20].

The inadvertent exposure of humans and wildlife to synthetic estrogens is a phenomenon of the last 60–70 years. Massive quantities of the pesticide DDT were first released into the environment at the start of World War II, with no knowledge of this chemical's estrogen mimicking ability [21] that was first revealed in 1950 [22]. In 1962 the public was alerted to the deleterious effects of pesticides on humans and wildlife with the publication of the book *Silent Spring* [23]. In response, action was taken to ban some chemicals including DDT. In the 1970s, as environmental exposure to these chemicals decreased significantly, the incidence of the most obvious toxic effects such as eggshell thinning decreased and the more subtle effects of these chemicals became apparent. Developmental and reproductive abnormalities have been reported in a wide variety of animal species. Notable examples are the feminization of male fish in sewage-fed rivers in Britain [24], of birds exposed in ovo [25], and of male alligators in Lake Apopka, Florida [26].

Generally, in cases of occupational exposure to adults, the deleterious reproductive effects are reversed once exposure to the estrogenic source is removed. However, research on in ovo exposure of birds to DDT [27] and in utero exposure of rodents and humans to DES [17, 28–30] indicates that the developing embryo or fetus suffers irreversible damage.

2 Alkylphenols

2.1 Identification of Estrogenic Activity

In 1991, nonylphenol was identified as a contaminant that leached from laboratory plasticware during normal use [31]. This compound was shown to be estrogenic using the E-SCREEN assay that measures estrogen-induced cell proliferation [32]. Further, nonylphenol was found to induce progesterone receptor (PR), which is also a marker of estrogenicity, and to induce mitotic activity in the rat endometrium [31].

2.2
Production and Uses

2.2.1
Alkylphenol Polyethoxylates

Reports in 1991 show that 450,000,000 pounds of alkylphenol polyethoxylates were sold in the U.S. during that year [33]. Produced from alkylphenols, these compounds comprise two major groups, nonylphenol polyethoxylate and octylphenol polyethoxylate. Nonylphenol polyethoxylate is the most prevalent, representing 80% of all alkylphenol polyethoxylates. Octylphenol polyethoxylate makes up most of the remaining 20% of alkylphenol polyethoxylates. Although not estrogenic themselves, these compounds degrade to alkylphenols that are estrogenic [31] during sewage treatment [34]. Currently, there is no standard for regulation of alkylphenol polyethoxylates in the United States.

Alkylphenol polyethoxylates have a wide variety of applications from industrial and institutional to household usage. Industrial applications comprise 55% of total use, institutional cleaners comprise 30%, and household cleaners and personal products comprise 15%.

2.2.1.1
Industrial Uses

Extensive reviews of the four major uses for alkylphenol polyethoxylates are provided by Talmage [35] and Dickey [36]. The first of these uses is in the production of plastics and elastomers. In this application they facilitate the polymerization process of acrylic and some vinyl acetate products and act to stabilize the final latex [35].

In the textile industry, alkylphenol polyethoxylates are used for cleaning fibers, scouring wool, and as finishing agents. Due to their good handling and rinsing characteristics, they are applied as wetting agents for spinning and weaving [35, 36].

Alkylphenol polyethoxylates are applied in agriculture as emulsifiers in the production of liquid pesticides, as wetting agents to enhance the adhesion of active chemicals to target organisms, and to facilitate spraying [35]. Of the registered pesticides, 4195 contain alkylphenol polyethoxylates, of which 81% are nonylphenol polyethoxylates and 19% are octylphenol polyethoxylates [36]. However, these compounds are often not included in the ingredient list of numerous products since they are classified as inert chemicals.

Finally, alkylphenol polyethoxylates are incorporated extensively in the production of paper and for recycling. They are used in pulping, which is the process to dissolve fibers, and for de-inking which is necessary for recycling print material [35].

2.2.1.2
Institutional Uses

Within this application, alkylphenol polyethoxylates are contained primarily in cleaners, with approximately 90% being used as commercial vehicle and metal cleaners. In addition, they act as non-chlorine sanitizers, deodorizers, and degreasers and are often found in both floor care and commercial laundry products.

2.2.1.3
Household Uses

There is an overlap in the use of alkylphenol polyethoxylates between household and industrial purposes, although the scale is different. In addition to their inclusion in laundry detergents and hard surface cleaners, they are present in personal care products such as shampoos, conditioners, hair coloring, and styling aids. The alkylphenol polyethoxylate, nonoxynol, is used as an effective spermicide in contraceptive creams and jellies, and prophylactics.

2.3
Human Exposure

Alkylphenol polyethoxylates are used in large volumes, thus providing significant potential for their release into the environment. Currently, little or no data are available on the release of these compounds into the atmosphere or on the contamination of products used for human consumption. Possible routes of exposure to alkylphenol polyethoxylates via air are from manufacturing plants and from the dispersal of pesticides during spraying. The direct exposure of humans to these compounds may occur by ingestion of pesticides on fruit and vegetables, consumption of fish that have bioaccumulated alkylphenol polyethoxylate metabolites through residence in contaminated waters, and via leaching of compounds into food and beverages from plastic storage containers. Research has shown that PVC tubing used in the processing of milk [37] and plastics used in food packaging [38] leach nonylphenol. Probably the main source of human exposure is from the use of nonoxynol-containing spermicides. Evidence in rodents shows that alkylphenol polyethoxylates, similar in nature to those used as spermicides, degrade to free nonylphenol [39].

2.4
Environmental Release

The majority of information available on the environmental presence of alkylphenol polyethoxylates and their degradation products is from the analysis of municipal wastewater treatment plants, rivers, and other water samples. The main source of alkylphenol polyethoxylates in water is from sewage discharges. Within the United States, a survey of drinking water in New Jersey revealed that a variety of alkylphenol polyethoxylates and their derivatives were present at a concentration of approximately 25 ng/l [40]. These compounds were also found in the wastewater, groundwater, and sewage on Cape Cod, Massachusetts, and both nonylphenol tetraethoxylate and octylphenol tetraethoxylate were found in one drinking water well at a concentration of 32.9 μg/l [41].

The most convincing indication of water contamination by estrogenic compounds has been in the United Kingdom. Reports by anglers of hermaphroditic fish triggered an investigation into possible sources of estrogenic contamination in rivers. It was found that the placement of male fish in various locations downstream of wool mills induced them to produce the female egg protein vitellogenin, which is normally produced in response to endogenous estrogen. The causal agents were found to be a mixture of estrogenic chemicals that were present within the effluent, the most prevalent being alkylphenol polyethoxylates. This effluent constituted as much as 80–90% of the total river flow. The estrogenic effect on fish was confirmed when some of the wool mills voluntarily ceased using alkylphenol ethoxylates. This coincided with a decrease in the level of vitellogenin production in the male fish (see chapter by Sumpter: Endocrine Disruptors in the Aquatic Environment, Part II).

A number of surveys have been conducted to determine the levels of alkylphenols and alkylphenol polyethoxylates in water sources throughout Europe and the United States. Levels of these compounds may fluctuate depending on the size of the river, the amount of rainfall, the amount of outflow from sewage treatment, and the microbial content of the river [42]. The results for surveys conducted in United States waterways of nonylphenol, nonylphenol polyethoxylate, octylphenol, and octylphenol polyethoxylate are summarized in Table 1.

As is evident from Table 1, there is a wide variation in the amounts of alkylphenol and alkylphenol polyethoxylates found in the environment. A more extensive survey needs to be undertaken to determine the magnitude of contamination.

2.5 Biodegradation

The wide range of uses for alkylphenol polyethoxylates results in its accumulation in industrial discharges, septic tanks, and sewage treatment plants. It is at these sites, and in the environment, that degradation by microbial action takes place. During this process, the polyethoxylate chain is shortened, producing free alkylphenols, or monoethoxylates and diethoxylates. Carboxylation may occur at the terminal end of these ethoxylate chains. The rate of degradation is influenced by the structure of the alkylphenol moiety and the length of the ethoxylate chain. Alkylphenols with a linear alkyl chain are more biodegradable than those with a branched chain, and *para* alkylphenols are more degradable than *meta* and *ortho* alkylphenols. Similarly, long chains take more time to degrade than shorter chains [44].

Degradation of alkylphenol polyethoxylates occurs through either anaerobic or aerobic pathways. Anaerobic degradation results in the formation of free

Table 1. Surveys conducted in United States waterways of nonylphenol, nonylphenol polyethoxylate, octylphenol, and octylphenol polyethoxylate

		Concentration (μg/l)	Highest concentration detected (μg/l)
NP	U.S. River Survey[a]	0.12	0.64
	Cape Cod[b]	29	33
NPE	U.S. River Survey	0.09	0.60
	Cape Cod	18	21
	New Jersey[c]	0.111	
NPE2	U.S. River Survey	0.10	1.20
	Cape Cod	7.2	8
	New Jersey	0.113	
NPE3–17	U.S. River Survey	2.0	14.9
NPE3–7	New Jersey	0.501	
OP	Cape Cod	0.47	0.74
OP2	New Jersey	0.032	
	Cape Cod	0.067	
OP3–8	New Jersey	0.124	

NP = nonylphenol, NPE = nonylphenol monoethoxylate, NPE2 = nonylphenol diethoxylate, NPE3–17=nonylphenyl polyethoxylates where the carbon chain equals 3–17, OP = octylphenol, OP2 = octylphenol diethoxylate, OP3–8 = octylphenol polyethoxylates where the carbon chain equals 3–8.

[a] Talmage [43].

[b] Clark et al. [40].

[c] Rudel et al. [41].

alkylphenols [34]. Smaller metabolites, such as nonylphenol, nonylphenol monoethoxylate, and nonylphenol diethoxylate, are more persistent in the environment and have been found in secondary effluents, sludge and digested sludge [43]. Both free alkylphenols and alkylphenol diethoxylates are estrogenic in character [45].

In aerobic degradation, the final products are water and carbon dioxide, this process being termed mineralization or ultimate biodegradation [46]. However, there is little evidence of this process occurring in alkylphenol polyethoxylates, in particular of the way in which the phenol ring is broken or the alkyl chain is degraded [47]. In the words of Ahel and colleagues, "if existing laws that mandate 80% biodegradability for surfactants are interpreted in terms of ultimate degradation to carbon dioxide and water, the alkylphenol polyethoxylates do not fulfill this basic requirement for environmental acceptability" [48].

Degradation of alkylphenol polyethoxylates within treatment plants is limited. A study of 16 wastewater treatment plants in Canada measured levels of 4-nonylphenol, nonylphenol monoethoxylate, nonylphenol diethoxylate, and 4-*tert*-octylphenol in raw sewage, sludge and effluent compartments. Of the 16 plants sampled, all contained measurable quantities of these phenolic compounds in raw sewage and sludge as expected. However, octylphenol was also present in effluent from 10 of the plants, and nonylphenol, nonylphenol monoethoxylate, and nonylphenol diethoxylate were present in effluent from almost all 16 sites [49].

The amount of alkylphenol polyethoxylates removed at wastewater treatment plants is debatable. One report suggests that 40% of incoming nonylphenol polyethoxylates are removed following treatment [48], while an industry-sponsored study [50] claims that greater than 95% is removed. However, it should be noted that the term "removal" does not describe the complete transformation of compounds to carbon dioxide and water. Due to their reduced polarity, free alkylphenols are adsorbed to the sludge within treatment systems and are therefore re-released into the environment in an undegraded, concentrated state. Reports of adsorption/partition coefficients for octylphenol polyethoxylate ($n=11$) and nonylphenol polyethoxylate ($n=10$) indicate up to a 1400-fold and 7500-fold concentration of these compounds, respectively, within sludge [51].

Alkylphenol polyethoxylates contained within sludge may persist in the environment. Reports indicate that sludge transferred to landfills under anaerobic conditions show practically no biodegradation of nonylphenol nor nonylphenol monoethoxylate over a 15-year period [52]. In semiaerobic conditions, degradation was over 90%. There is some suggestion that sludge discharged to sand beds may percolate and contaminate groundwater [53].

Nonylphenol polyethoxylates degrade naturally within the environment to form a number of persistent metabolites. A study of two field sites in northern Switzerland revealed high levels of nonylphenol, nonylphenol monoethoxylate, nonylphenol diethoxylate, and nonylphenoxycarboxylic acids in rivers. Significantly lower levels of nonylphenol, nonylphenol monoethoxylate, and nonylphenol diethoxylate were found in groundwater [54], suggesting that percolation through soil may eliminate these compounds.

Finally, in addition to their biodegradation and adsorption to sediment, nonylphenol and octylphenol migrate from surface waters to the air due to their volatility [55].

2.6 Bioaccumulation and Metabolism

The accumulation of alkylphenols within aquatic and terrestrial environments increases their potential to be incorporated into the food cycle. By virtue of their lipophilic nature, alkylphenols bioaccumulate within adipose tissue of fish and mammals [56] and therefore are available for consumption by humans. Vertebrates are unable to degrade the phenol ring, and thus the cycle of environmental exposure to alkylphenols is perpetuated by the release of unmetabolized compounds from human and animal waste back into the environment.

The bioaccumulation and metabolism of alkylphenols within rainbow trout have been studied by exposing them to radiolabeled nonylphenol within tanks. The ratio of radioactivity per gram in the tank water relative to tissues in the trout (bioaccumulation factor) revealed that nonylphenol accumulates predominantly within the viscera (bioaccumulation factor of 98 relative to that of 24 in the carcass) with greatest abundance being in liver, fat, kidneys, and bile [57]. This study may underestimate true aquatic bioaccumulation since it was performed in a static system, in which a finite amount of nonylphenol was available, rather than in a flow-through system in which nonylphenol would be continuously present.

These findings concur with other studies in fish. Exposure of rainbow trout to [^{3}H]-4-nonylphenol revealed accumulation of the parent compound in muscle and the presence of metabolites, specifically glucuronic acid conjugates, in liver, bile, and feces. The half-life of the radiolabeled nonylphenol was 99 h and 97 h in edible muscle and skin, respectively, and 5 h in liver [58]. Although significantly less than the half-life of PCBs (polychlorinated biphenyls), which is years, these data for nonylphenol are not negligible when environmental exposure occurs daily. Comparable results have been documented using ^{14}C-nonylphenol [57]. These simulated studies are supported by environmental evidence. The level of alkylphenols in carp residing in the Detroit River, Michigan, are two- to sixfold greater than in the surrounding sediment, indicating bioaccumulation in fat [59].

It has been demonstrated that nonylphenol polyethoxylates are metabolized in rats [39]. The administration of alkylphenol polyethoxylates, ^{14}C-labeled in the ethoxylate chain, results in excretion of ^{14}C-labeled nonylphenol polyethoxylate in the feces and urine, and ^{14}C-labeled carbon dioxide through the lungs. The alkylphenols found in urine were conjugated. In this study, alkylphenol polyethoxylates labeled in the phenol ring never produced ^{14}C-labeled carbon dioxide. After 4 days, approximately 90% of the combined nonylphenols were excreted.

2.7 Biological Effects

There is evidence to suggest that high doses of alkylphenol derivatives have profound effects on the development of rodents. Oral administration of 250 mg/kg and 500 mg/kg nonoxynol-9 per day to pregnant rats have been shown to induce a significant decrease in maternal weight gain, and an increased incidence of developmental abnormalities in offspring such as extra ribs and dilated pelvises [60].

2.8 Developmental and Reproductive Effects

As described in Sect. 2.4, investigators revealed that male fish residing in certain rivers downstream of wool mills in the United Kingdom produced vitellogenin in response to estrogenic contamination by effluent [61]. Subsequent experiments revealed the presence of alkylphenols in the sewage effluent. These chemicals were tested individually for their ability to induce vitellogenin production in rainbow trout and in primary cultures of rainbow trout hepatocytes; 4-nonylphenol, 4-nonylphenoxycarboxylic acid, nonylphenol diethoxylate, and 4-*tert*-octylphenol all induced vitellogenin production in a dose-dependent manner. Octylphenol was shown to be the most potent [45]. The chemical 4-*tert*-octylphenol caused inhibition of testicular growth by 50%; 4-nonylphenol, 4-nonylphenoxy carboxylic acid, and nonylphenol diethoxylate had the same effect although to a lesser degree [62].

Purdom and colleagues demonstrated that other species of fish are affected similarly by these chemicals. Measurements in carp revealed an increase in the production of vitellogenin [61]. Juvenile Atlantic salmon showed variations in steroid hydroxylases, cytochrome P450 isozymes, and conjugating enzyme levels caused by nonylphenol [63]. Exposure of Japanese medaka to 50 µg/l and 100 µg/l nonylphenol from the time of hatching to 3 months of age caused some of the fish to develop ovotestis and an alteration in the ratio of males to females [64].

The adverse effects of alkylphenols on development and reproduction have also been documented in mammals. Nonylphenol has been shown to induce cell proliferation in the luminal epithelium of the endometrium in ovariectomized rats [65]. Studies in males demonstrated that exposure of neonatal rats to octylphenol (6 doses of 2 mg over 12 days) caused a reduction in testicular weight by adulthood, although no change was evident at day 18 [66]. Similarly, 0.8 mg/kg nonylphenol administered to rats caused a dose-dependent decrease in the relative weights of the testis, epididymis, seminal vesicle, and ventral prostate. The effects described for this latter study were time-dependent and were evident in pups dosed before 13 days of age but not subsequent to this period [67].

A study by Sharpe et al. [68] demonstrated that a dose of 1000 µg/l octylphenol and octylphenol polyethoxylate administered to rats through their drinking water induced a small but significant decrease in testicular weight and a 10–21% reduction in daily sperm production. No changes in testicular morphology were evident. However, subsequent attempts made by the same authors to repeat this study were unsuccessful [69].

3
Bisphenol A

3.1
History

The first phenolic plastic to be manufactured for use in industrial and household applications was patented under the name Bakelite in 1909 [70]. It was not until 1993 when Krishnan and colleagues [71] published a serendipitous discovery made while seeking evidence of estrogen production in yeast, that attention was focused on the estrogenicity of phenolic resins. They determined that a substance, subsequently identified as bisphenol A (BPA, 2,2-bis(*p*-hydroxyphenyl)-propane), leached from laboratory polycarbonate flasks when autoclaved. This compound was confirmed to be estrogenic by its ability to bind the ER with an affinity approximately 1:2000 that of 17β-estradiol, upregulate the expression of PR, and induce cell proliferation in estrogen-sensitive MCF-7 cells cultured in vitro. Almost 60 years earlier BPA had been found to exhibit an estrogenic character [72]; however little was made then of this warning in the context of possible deleterious effects on organisms that came in contact with this chemical. Currently, BPA is the chemical used most widely in the manufacture of phenolic resins.

3.2 Identification of Estrogenic Activity

The estrogenicity of BPA has been demonstrated in a variety of in vitro and in vivo assays. It has been shown to induce cell proliferation in MCF-7 cells [71, 73, 74], stimulate release of prolactin from pituitary GH_3 cells [75], and induce transcriptional activation of ER in both yeast-based assays [76] and in human embryonal kidney cells via the estrogen response element (ERE) [77]. In addition, BPA has been shown to upregulate the expression of vitellogenin mRNA in primary hepatocytes from the male *Xenopus laevis* [78]. BPA binds both ER-α and β to an equal degree [77].

Our understanding of the true biological significance of exposure to BPA is currently hindered by a lack of information on the pharmacokinetics and pharmacodynamics of this compound. In vitro assays such as the E-SCREEN measure the target tissue doses of xenoestrogens relative to 17β-estradiol. Although these accurately reflect ER binding [74, 79] they show BPA to be significantly less potent than 17β-estradiol, as they do not take into account factors such as uptake, transportation, and metabolism specific to the live animal. As a result, in vitro assays overestimate the potency of 17β-estradiol relative to BPA and other xenoestrogens [80].

In the circulation, estradiol is bound to various serum proteins including albumin, sex hormone-binding globulin, corticosteroid-binding globulin, and α-fetoprotein that act as important modulators of endogenous hormone activity [81–83]. The binding affinity of endogenous estrogens and xenoestrogens to α-fetoprotein is of particular importance in rodent fetuses and neonates. This protein is believed to prevent early exposure of the developing organism to endogenous, natural estrogens, thus inhibiting inappropriate sexual differentiation of the brain [84]. Studies on the binding affinity of BPA to various serum proteins have demonstrated negligible binding for rat α-fetoprotein and a low binding affinity for human sex steroid-binding protein (0.01%) and trout sex steroid-binding protein (0.1%) relative to [^{3}H]dihydrotestosterone (DHT) [85]. Hence, this low affinity for plasma sex steroid-binding proteins may increase the effective concentration of BPA in circulation, make it more readily available to the ER, and thus enhance its estrogenic activity relative to the protein-bound estradiol.

Other bisphenols, such as bisphenol F, bisphenol AF, and additional diphenylalkanes, exhibit estrogenic properties. A correlation between structure and activity exists for these compounds such that the longer the alkyl substituents at the bridging carbon, the higher the estrogenic activity [70].

3.3 Production and Use

3.3.1 *Industrial Production Levels*

BPA is a monomer used in the production of polycarbonates and epoxy resins from which a wide variety of products are generated. These include automotive lenses, optical lenses, food and beverage containers, protective coatings, adhesives, powder paints, protective window glazing, building materials, compact disks, thermal paper, paper coatings, as a developer in dyes, and for the encapsulation of electrical and electronic parts [86]. Figures from 1995 show that BPA was one of the top 50 chemicals manufactured in the United States, with an output of over 1.6 billion pounds [87].

3.3.2 *Food and Beverage Containers*

Epoxy resins are used to lacquer-coat the interior of food cans, wine storage vats, water containers, and water pipes. Polycarbonate plastics are used to manufacture water carboys, reusable milk containers, food storage vessels, and babies formula bottles. It has been determined that incomplete polymerization of these products during manufacture and increased temperatures imposed during heating cause unreacted compounds to leach into foods and beverages [88]. BPA has been identified in the liquid within which canned vegetables are stored in levels ranging from 0–23 µg/can, depending upon the vegetable tested [73]. The highest of these concentrations was sufficient to induce in vitro proliferation of MCF-7 cells. Extracts from autoclaved cans that contained fatty foods also induced cell proliferation. It has been demonstrated that the practice of sterilizing plastic babies bottles and cups causes leaching of 7–58 µg/g BPA [89]. In addition, both BPA and bisphenol A diglycidyl ether (BADGE) have been identified in wine, presumably due to contact with epoxy during storage in vats [90]. Reports from a collaborative effort by the Society of Plastics, Keller and Heckman, and the National Food Processors Association state that BPA was found to be undetectable in extracts from beverage cans, but ranged from 0 to 121 parts per billion (ppb) in extracts from some food cans. The highest levels of BPA were extracted from infant formula and fruit juice cans [91].

3.3.3 *Dental Sealants and Composites*

The introduction of BPA polymers into dental restorative products occurred in response to problems associated with the original chemically-cured methacrylates. Thus, methacrylate resins containing bisphenol A-diglycerolmethacrylate (BisGMA) were introduced, proving to be superior due to their greater retention and ability to form strong cross-links. Triethyleneglycol dimethacrylate (TEGDMA) was added also to reduce the high viscosity inherent in these products and to enhance their manipulative properties. These compounds are the constituents of most commercially available dental sealants, which are used as preventive, anti-caries coatings on teeth, and composites, which are white fillings. An extensive review of this literature is provided by Soderholm and Mariotti [92].

These sealants are polymerized by either ultraviolet light or by interaction with a benzoyl peroxide initiator with a tertiary amine activator. Strong evidence has emerged to suggest that the polymerization process is not always complete and that BPA and associated polymers leach into the saliva of patients [93]. Furthermore, enzymatic hydrolysis of methacrylates and mechanical forces upon the tooth surface contribute to the persistent degradation of dental resins. In a study that has generated much disquiet in the dental industry, Olea and colleagues demonstrated the presence of BPA, in addition to other BPA derivatives (bisphenol A-dimethacrylate (BPA-DMA), BisGMA, and BADGE), in the saliva of patients. The saliva samples, which were collected over a 1-h period, were found to contain 90–931 µg following the application of 50 mg of a commercial BisGMA-based dental sealant to the patients' molar teeth. The saliva concentrations of BPA and BPA-DMA were sufficient to induce cell proliferation in MCF-7 cells and increase both PR and pS2 levels, confirming the compounds' estrogenic activity. One patient exhibited residual BPA and BPA-DMA in her saliva from dental work performed 2 years previously, countering arguments that leakage of BPA products only occurs immediately following the dental procedure.

Recent studies have confirmed the leakage of TEGDMA, BPA-DMA, and BisGMA from a group of commercially available sealants; however, they question the presence of BPA shown in Olea's study [94, 95]. This discrepancy may be due to differences in the experimental protocol. The latter investigations were performed in vitro and the leachants measured in distilled water and ethanol washes rather than saliva. None of these studies tested the estrogenic activity of the leached compounds. However, one study demonstrated that a dose of 100 mg/kg BisGMA administered subcutaneously to ovariectomized mice (3 times per week for 3 weeks) induced an increase in uterine wet weight and collagen content [96]. Currently, the European Union has established a specific migration limit of 3 mg/kg for BPA and 0.02 mg/kg for BADGE. The European Committee for Food has estimated a daily intake of BPA at 0.05 mg/kg body weight.

3.3.4 *Medical Materials*

The use of bioactive bone cement containing BisGMA was introduced into orthopedic medicine for applications such as osseous repair of the skull cap, reconstruction of the anterior wall of the frontal sinus, and fixation of alloy implants [97, 98]. Combined with apatite-wollastonite glass ceramic, BisGMA-based resin provides an effective means of fixing hip prostheses, the composition of which has been researched in rats [99, 100], rabbits [101], and dogs [102]. This material's superior mechanical strength, ability to bond directly to bone, and proposed good bioactivity make it an advantageous alternative to conventional materials such as polymethyl methacrylate cement. The issue of potential leakage of partially polymerized BPA from orthopedic materials and the possible effects has not been addressed in the literature. One study, describing the advantages of BisGMA/apatite-wollastonite glass ceramic cement, expressed concern that these materials with high bioactivity are often less stable and subject to severe biodegradation [100].

3.4 Human Exposure

The application of dental sealants and composites to teeth and their subsequent degradation by enzymatic and mechanical forces may result in BPA and its derivatives being absorbed through the gingival epithelium or being swallowed. Evidence in mice suggests that these products enter the gastrointestinal tract and may be absorbed through its epithelium [103]. Dental personnel may be exposed to BPA and its derivatives through skin contact in the preparation and application of restorative dental materials. In manufacturing facilities, workers are generally exposed to large doses of BPA through inhalation and skin contact; the latter has been shown to cause photosensitive dermatitis [104]. In homes, exposure to unhardened epoxy resins containing BPA occurs predominantly through skin contact with coated household objects and hobby glues.

3.5 Environmental Release

The figures in 1996 for total environmental release of BPA were approximately 465,000 pounds. Of that, 39.5% comprised total air release, 1% total water release, 54% total land release, and 5.4% total underground injection (NIH, 1998: http://toxnet.nlm.nih.gov/servlets/). This environmental waste is most likely generated in the manufacturing process and released during processing, handling, and transportation. Of the approximate 1.6 billion pounds produced, the remaining portion of BPA is present in polycarbonate and epoxy resin products previously outlined. Recent studies in Cape Cod, Massachusetts, measured BPA levels of 0.1 – 1.7 µg/l in untreated septic- and wastewater and 20–44 ng/l in 2 of 28 drinking water wells [41].

3.6 Bioaccumulation

Xenoestrogens are generally lipophilic, a characteristic that may facilitate their absorption through skin and mucous membranes, and accumulation in food-producing animals. Elimination profiles of a single dose of ^{14}C-labeled BPA administered orally to male CFE rats revealed that 56% was excreted in feces and 28% was excreted in the urine, indicating absorption through the intestinal wall. After 8 days of a single exposure, there were no traces of radiolabeled BPA left in the body [105]. In CF-1 mice, elimination profiles of a single dermal application of ^{14}C-labeled BADGE showed that 20% was excreted in feces, 3% was excreted in urine, and 66% was extracted from the skin at the area of application over 3 days [103]. When BADGE was administered orally in the same study, 80% was excreted in feces and 11% was excreted in urine over 3 days. Within 2 days, 88% of the total administered dose was excreted and by 8 days only 0.1% of the dose remained. Recent studies in pregnant Fischer rats have demonstrated that BPA, administered in a single oral dose, is able to rapidly traverse the placenta and distribute within fetal organs. Following a single oral dose of 1 g/kg BPA to rats, the chemical was found to reach a maximal concentration within fetal organs by 20 minutes; after 40 minutes the concentration of BPA was higher in the fetus than in the maternal blood [106].

In a collaborative effort by a consortium of companies including Shell Development Company, Dow Chemical Company, and Society of the Plastics Industry, Staples and colleagues provide an extensive review of the environmental fate, bioaccumulation, and biodegradation of BPA [86]. In the aquatic models presented, the authors claim that there is a low potential for BPA to bioaccumulate in microorganisms, algae, invertebrates, and both freshwater and marine fish. No other recent study is available to challenge or confirm these assertions.

3.7 Metabolism

There is a paucity of information on the metabolism of BPA and its derivatives in animals and humans. Studies show that BPA is metabolized to hydroxylated BPA [105] and then oxidized to form an ortho-quinone. This latter product is capable of covalently binding with DNA [107], indicating the potential for BPA to modify DNA nucleotides and cause mutational change. Research in CD-1 rats demonstrated that BPA produced covalent modifications in testicular DNA in vivo [108]. Studies of BADGE show that the major metabolic pathway of this BPA derivative is the hydrolytic opening of the two epoxide groups to form the bis-diol of BADGE. This compound is excreted in both free and conjugated forms, although the majority undergoes further metabolic transformations to form carboxylic acids [109].

3.8 Biodegradation

The review of mostly environmental models by Staples and colleagues [86] concludes that BPA generally undergoes rapid biodegradation in surface waters, wastewater treatment plants, and biological waste treatment systems at an efficiency of greater than 96%. The short half-life of BPA in test effluent (2.5–4 days), rapid acclimation of microbial populations to degrade this compound, and the potential for photo-oxidation are factors that facilitate its rapid biodegradation in the environment by mineralization. In contrast, studies by Stone and Watkinson [110] found insufficient evidence to conclude that BPA is readily biodegraded.

Research on the bacterial biodegradation pathway of BPA reveals that the metabolites 4-hydroxyacetophenone and 4-hydroxybenzoic acid are formed in the major pathway, both of which are rapidly degraded to carbon dioxide and water. The minor pathway of BPA degradation produces the metabolites 2,2-bis(4-hydroxyphenyl)-1-propanol and 2,3-bis(4-hydroxyphenyl)-1,2-propanediol that could further degrade to form carbon dioxide, be incorporated in bacterial cell material, or form miscellaneous soluble organic compounds [111].

3.9 Biological Effects

The various reports on the biological effects of BPA are quite diverse since the premises upon which the investigations are based differ significantly. Toxicology studies aim to induce a pronounced biological effect and thus large doses of BPA have been administered in animal models. In contrast, studies in which much lower doses are administered to animals and to in vitro systems question the estrogenic potency of BPA. These investigations are based upon a rationale that aims to understand the effects of BPA on the development and reproduction of animals, and ultimately humans, at physiologically relevant doses.

3.9.1 *Teratogenic Effects*

The in utero exposure of Sprague-Dawley rats to 85–125 mg/kg BPA (gestational days 1–15) have been shown to cause imperforate anus, incomplete skeletal ossification, and enlarged cerebral ventricles [112]. These levels also cause maternal toxicity, a reduction in the number of pregnant rats, and a reduction in the number of live fetuses born. In CD rats, doses of 160–640 mg/kg/day delivered orally caused a decrease in maternal weight only. When similar experiments were performed on CD-1 mice, doses of 500–1250 mg/kg/day caused an increase in maternal liver weight and maternal mortality [113]. The highest dose caused an increase in the resorption rate of fetuses exposed in utero per litter, and a decrease in both fetal body weight and uterine weight. Alterations in the gross fetal developmental of either species were not evident.

3.9.2 *Developmental Effects*

Exposure of the developing male fetus to estrogens causes significant changes in the reproductive tract in adulthood. The in utero positioning of a male mouse between 2 female siblings is associated with a 20% increase in prostate weight [114], a phenomenon that can be reproduced by exposing the developing male fetus to a minor increase in serum estradiol [115]. BPA has the capacity to induce the same effect. In utero exposure of CF-1 male mice to 2 µg/kg and 20 µg/kg BPA was found to cause an increase in adult prostate weight by 30% and 35%, respectively. Body weight decreased in response to BPA exposure, this effect being significant only with the low dose [82]. These findings are compatible with research revealing that changes in prostate weight follow an inverted U curve in response to increasing in utero doses of both 17β-estradiol and DES [115]. These same doses caused a permanent increase in the preputial gland size and a decrease in epididymal size in males exposed in utero; 20 µg/kg BPA caused a decrease in sperm production by 20% [116]. The significance of these data is of extreme importance considering that the lower of these doses is less than that reported in saliva after administration of dental sealants [93].

These data confirm the inappropriateness of assuming that, in experimental conditions, BPA is being administered to estrogen-naïve animals. Fetuses are already exposed to endogenous estrogens. In view of the results showing a non-monotonic dose response to BPA, it should be considered that there is no concept of a threshold for hormone mimics [117].

BPA has been shown to cause advancement in the onset of puberty in female CF-1 mice. In utero exposure of mice to 2.4 µg/kg BPA between gestational days 11 and 17 caused a decreased survival of female pups between birth and weaning, an increase in pup weight at weaning, and a decrease in the number of days between vaginal opening and first uterine estrus [118]. These effects were enhanced if females were positioned between either 1 or 2 female siblings in utero.

Exposure of Sprague-Dawley female rats to 100 µg BPA/kg body weight/day throughout in utero development and lactation has been shown to cause a significant increase in body weight at birth. This increase persisted throughout the observation period which extended to 110 days. A tenfold higher dose resulted in a significant increase in body weight over controls for the first two weeks [119]. These data are consistent with other observations of non-monotonic dose-response curves [115, 120, 121]. In addition, exposure of rats to the high dose of BPA resulted in a decreased number of animals showing estrous cyclicity at 4 months of age relative to controls.

Recent work in our laboratory has revealed that *in utero* exposure of CD-1 mice to low, presumably environmentally-relevant doses of BPA (25 and 250 µg/kg) induced changes in the developmental timing of DNA synthesis within the epithelium and stroma of the mammary gland. In 6 month-old mice, the histoarchitecture of the mammary glands resembled those of early pregnancy (increased presence of all epithelial structures, including an approximate 300% increase in alveolar buds; greater presence of secretory product), and suggested that prenatal exposure to BPA may predispose the tissue to neoplastic change in adulthood [122].

Studies in male Wistar rats have demonstrated that exposure to a 0.5 mg in 20 µl injection of BPA on days 2, 4, 6, 8, 10, and 12 of neonatal life caused no changes in testis weight or seminiferous tubule diameter. Similarly, the immunoexpression of follicle stimulating hormone-β (FSH-β) in pituitaries and inhibin α-subunit in Sertoli and Leydig cells remained unaltered [66].

It has been shown that BPA has dramatic effects on amphibia as well as on mammals. The exposure of *Xenopus laevis* to 10^{-7} mol/l BPA induced feminization of sexual differentiation. This effect was evident by a significant increase in female phenotypes of larvae compared to controls [78].

3.9.3
Reproductive Effects

The proliferative effect of natural estrogens on the uterus has traditionally been the hallmark of estrogen action. This response has formed the basis of the mouse uterotropic assay [123]. The assay detects the estrogenicity of a suspected compound by its ability to induce an increase in wet weight of a prepubertal mouse uterus within 3 days [124]. The studies reviewed below reveal that there is a wide variety of sensitivity to BPA among species and strains of rodents.

Studies in immature CFLP mice showed that doses of 0.05 mg and 0.5 mg per mouse (assuming a mouse weight of approximately 30 g, this represents a dose of 1.67 mg/kg and 16.7 mg/kg) were insufficient to induce a uterotropic effect. Yet a 5 mg (167 mg/kg) dose caused toxicity to the animal [125]. Work in our laboratory has established a dose-response curve for BPA in the CD-1 mouse, and has revealed that for the parameters of age of vaginal opening and uterine wet weight, a non-monotonic dose response exists for this chemical. That is, the lower (0.1 mg/kg) and higher doses (75 and 100 mg/kg) of BPA induce significant changes in these reproductive parameters, while the middle range of doses have no effect [126].

Immature Alpk:AP rats treated for 3 days with a dose of 400 mg/kg BPA administered by either oral gavage or subcutaneous injection showed a significant increase in uterine wet weight [127]. These findings are consistent with those of Dodds and Lawson [72], who showed that a total of 100 mg BPA injected twice daily for 3 days in ovariectomized rats (of an unstated weight) induced persistent estrus. Assuming a rat weight of 250 g, this represents a dose of 800 mg/kg/day. Yet another study revealed that a minimum daily dose of 10 mg/kg and 30 mg/kg BPA delivered orally for 4 days in ovariectomized Sprague Dawley rats induced a 29% and 37% increase in uterine wet weight, respectively [128].

Steinmetz and colleagues determined that a single 37.5 mg/kg dose of BPA administered intraperitoneally to Fischer rats induced a significant increase in bromodeoxyuridine (BrdU) labeling of vaginal and uterine epithelial cells 20 h later, indicating cell proliferation [129]. A 50 mg/kg dose of BPA induced an increase in the expression of c-*fos* mRNA in the luminal epithelium of the uterus by 14- to 17-fold and in the vagina by 7- to 9-fold within 2 h. Subcutaneous delivery of approximately 0.3 mg/kg BPA per day for 3 days caused a marginal increase in uterine wet weight and a 2.5-fold increase in luminal epithelial cell height and mucus secretion. Proliferation of the vaginal epithelium from 2–3 cell layers to 6–8 cell layers and cornification were also seen. These changes induced in Fischer 344 rats were not evident in Sprague Dawley rats at the same dose [129].

Research in Noble rats demonstrated that a 0.1 mg/kg/day and 54 mg/kg/day exposure to BPA for 11 days induced a 143% and 220% increase, respectively, in proliferative activity of mammary gland epithelium [130]. These changes were associated with a significant increase in the conversion of immature to mature glandular structures in both the low and high dose groups, indicating that low doses of BPA can induce profound proliferative effects in mammary glands.

Longitudinal studies in rodents suggest that BPA causes reproductive toxicity that persists into the second generation. One study of CD-1 mice revealed that exposure to high levels of BPA via ingestion (low dose: 437 mg/kg/day BPA; medium dose: 875 mg/kg/day; high dose: 1750 mg/kg/day) caused a longer gestation period and decreased litter size in the high dose range [131]. F1 females appeared to be the most affected as they delivered 51 % fewer pups when mated with control partners. The males sired 25 % fewer pups in the high BPA group. On dissection, both F0 and F1 generations exhibited an increase in liver and kidney weights. The males exhibited decreased seminal vesicle weight, with compromised sperm motility in the parents only. In the high dose F1 mice, pup mortality prior to weaning was significantly increased.

3.9.4
Neuroendocrine Axis

The neuroendocrine axis is an integral part of reproductive function. However, few studies have been undertaken to assess the response of this system to BPA exposure. One study on Sprague Dawley rats showed that in utero and lactational exposure to a minimum dose of 320 mg/kg/day induced an 85 % increase in the volume of the sexually dimorphic nucleus of the preoptic area in the brain. This effect was seen in neonatal females only [132]. In vitro work revealed a dose-dependent increase in the release of prolactin from anterior pituitary cells harvested from ovariectomized Fischer 344 rats in response to BPA [75]. In the same study, 1 nmol/l BPA induced a threefold increase in prolactin release from cultured GH_3 cells, a somatomammotroph cell line, by 5 days. These experiments translated well into the animal model, which demonstrated that an exposure of 40–45 µg/day BPA for 3 days induced a 7- to 8-fold increase in serum prolactin levels in Fischer 344 rats, but not in Sprague Dawley rats. Assuming a rat weight of 200 g, this represents a dose of 200–225 µg/kg/day.

4
Conclusion

A diverse number of chemicals that exhibit estrogenic activity are currently being used in large volumes. The potential for these chemicals to disrupt the development and normal functioning of organisms has justifiably provoked concern. The issue, therefore, is to determine whether the use of these endocrine disruptors should be regulated to curtail exposure of humans and wildlife.

The knowledge that alkylphenol polyethoxylates are toxic to aquatic organisms long preceded the findings that alkylphenols are estrogen mimics. Evidence became apparent when fishes residing downstream of industrial effluent outlets showed signs of estrogenicity. When the mills switched voluntarily from using alkylphenol polyethoxylates to non-estrogenic detergents (alkyl ethoxylates), the levels of alkylphenols and signs of estrogenic activity in the fishes declined concomitantly. There is little data on alkylphenol exposure to non-aquatic organisms including humans; however, it is reasonable to assume that aquatic species are the ones most likely to be affected. These species are exposed constantly to phenolic chemicals, while other species may be exposed only intermittently. Nonetheless, nonylphenol accumulates in the muscle of fish and therefore their predators are likely to be affected. The most obvious exposure of alkylphenols to humans is through the use of nonoxynol spermicides. The effects of such exposure have not been studied in detail as yet.

BPA and other phenolic compounds have been used in the manufacture of plastics since the introduction of Bakelite, which was patented in 1909. There is very little information concerning the levels of these compounds in the environment. Humans are probably the most exposed species, since BPA products are used in the food industry and as medical and dental materials.

In vitro studies testing the estrogenic potency of BPA have provided knowledge of the dose required to induce an effect at the target cell. However, such models are not representative of the fate of this chemical in the organism, and do not take into account metabolism, binding to plasma proteins and other pharmacokinetic issues. It is wrong to assume that because these compounds show a low potency relative to estrogen in vitro that they are harmless. This has motivated scientists to study the effects of BPA in rodents. The uterotropic assay, which is the classical tool for assessing estrogenicity, appears to be rather insensitive. When other parameters are considered, such as the induction of proliferative activity in the epithelia of the vagina, uterus, and mammary gland, tissue changes are observed at doses that are ineffective in the uterotropic assay. The administration of BPA during prenatal and early postnatal development induces effects at doses that are orders of magnitude lower than that needed for a positive uterotropic response. As we learn more about the unintended biological effect of alkylphenols and BPA, it becomes apparent that their effect is most striking and irreversible when exposure occurs during embryonic development.

In most toxicological studies, it is assumed that the dose-response curve is monotonic. It is believed that testing very high doses will suffice to assess all the effects of a chemical. However, there is evidence to suggest that sex steroids produce varied effects at different doses. Androgens, for example, induce proliferation of prostate epithelial cells at a relatively low dose, and inhibit cell proliferation at higher doses. Furthermore, the effects of estrogens on the developing genital tract follow an inverted U dose-response curve. These findings are revealing that our general assumptions are wrong. We must look specifically at low dose effects, that is, those that occur at actual levels of human exposure because testing at high doses may mask these effects.

This chapter has focused on the properties and biological effects of alkylphenols and BPA individually. Although this is an appropriate beginning, it is becoming evident that we must consider these chemicals as part of a mixture. It should be taken into consideration that wildlife and humans are not exposed to one single chemical at a time, and that hormone mimics are acting upon organisms that are already exposed to endogenous hormones and other xenobiotics. Thus, the other classical assumption that we need to reject is that of the existence of a threshold.

The findings of the last decades up until this point have brought us abruptly to face the facts that tampering with the ecosystem brings unforeseen consequences. Finally we must acknowledge our ignorance and temper our scientific arrogance. Our quest as scientists now is to look for new ways to study these complex systems.

5
References

1. Colborn T, vom Saal FS, Soto AM (1993) Environ Health Perspect 101:378
2. Kelce WR, Monosson E, Gamcsik MP, Laws SC, Gray LE Jr (1994) Toxicol Appl Pharmacol 126:276
3. Treinen KA, Heindel JJ (1992) Reprod Toxicol 6:143
4. Treinen KA, Dodson WC, Heindel JJ (1990) Toxicol Appl Pharmacol 106:334
5. Davis B, Maronpot R, Heindel JJ (1994) Toxicol Appl Pharmacol 128:216
6. Hertz, R (1985) The estrogen problem – retrospect and prospect. In: McLachlan JA (ed) Estrogens in the environment II – influences on development. Elsevier, New York, p 1
7. Masiakowski P, Breathnach R, Bloch J, Gannon F, Krust A, Chambon P (1982) Nucleic Acids Res 10:7897
8. Kuiper GG, Enmark E, Pelto-Huikko M, Nilsson S, Gustafsson JA (1996) Proc Natl Acad Sci USA 93:5925
9. Kuiper GG, Carlsson B, Grandien K, Enmark E, Haggblad J, Nilsson S, Gustafsson JA (1997) Endocrinology 138:863
10. Lubahn DB, Moyer JS, Golding TS, Couse JF, Korach KS, Smithies O (1993) Proc Natl Acad Sci USA 90:11,162
11. Bern HA (1992) The fragile fetus. In: Colburn T, Clement C (eds) Chemically-induced alterations in sexual and functional development: the wildlife/human connection. Princeton Scientific Publishing, Princeton, p 9
12. Newbold, RR, McLachlan, JA (1985) Diethylstilbestrol associated defects in murine genital tract development. In: McLachlan JA (ed) Estrogens in the environment II: influences on development. Elsevier Science Publishing, New York, p 288
13. Smith EP, Boyd J, Frank GR, Takahashi H, Cohen RM, Specker B, Williams TC, Lubahn DB, Korach KS (1994) New Engl J Med 331:1056
14. vom Saal FS, Grant WM, McMullen CW, Laves KS (1983) Science 220:1306
15. Finkelstein J, McCully W, MacLaughlin D, Godine J (1988) N Engl J Med 318:961
16. Mittendorf R (1995) Teratology 51:435
17. Herbst AL, Anderson D (1990) Semin Surg Oncol 6:343
18. Ingle JN, Ahman DL, Green SJ (1981) N Engl J Med 304:16
19. Boyer MJ, Tattersall MHN (1990) Med Pediat Oncol 18:317
20. Pitts WR Jr (1999) Urology 53:660
21. Thomas KB, Colborn T (1992) Organochlorine endocrine dispruptors in human tissue. In: Colborn T, Clement C (eds) Chemically induced alterations in sexual development: the wildlife/human connection. Princeton Scientific Publishing, Princeton, NJ, p 365
22. Burlington H, Lindeman VF (1950) Proc Soc Exp Biol Med 74:48
23. Carson R (1987) Silent spring: 25th anniversary edition. Houghton Mifflin, New York
24. Sumpter JP (1998) Arch Toxicol Suppl. 20:143
25. Fry DM, Toone CK (1981) Science 213:922
26. Guillette LJ, Gross TS, Masson GR, Matter JM, Percival HF, Woodward AR (1994) Environ Health Perspect 102:680
27. Fry DM (1987) Stud Avian Biol 10:26
28. Pylkkanen L, Santti R, Newbold RR, McLachlan JA (1991) Prostate 18:117
29. Newbold RR, Bullock BC, McLachlan JA (1990) Cancer Res 50:7677
30. Bern HA (1992) Diethylstilbestrol syndrome: present status of animal and human studies in hormonal carcinogenesis. Springer, Berlin Heidelberg New York
31. Soto AM, Justicia H, Wray JW, Sonnenschein C (1991) Environ Health Perspect 92:167
32. Soto, AM, Lin, T-M, Justicia, H, Silvia, RM, Sonnenschein, C (1992) An "in culture" bioassay to assess the estrogenicity of xenobiotics. In: Colborn T, Clement C (eds) Chemically induced alterations in sexual development: the wildlife/human connection. Princeton Scientific Publishing, Princeton NJ, p 295
33. The Chemical Manufacturers Association (1993) CMA Alkylphenol and Ethoxylates Panel

34. Giger W, Brunner PH, Schaffner C (1984) Science 225:623
35. Talmage SS (1994) Environmental and human safety of major surfactants. Lewis Publishers, Tokyo, p 200
36. Dickey P (1997) Troubling bubbles. Washington Toxics Coalition, Seattle
37. Junk GA, Svec HJ, Vick RD, Avery MJ (1974) Environ Sci Technol 8:1100
38. Gilbert MA, Shepherd MK, Startin JR, Wallwork MA (1992) J Chromatogr 237:249
39. Knaak JB, Elridge JM, Sullivan LJ (1966) Toxicol Appl Pharmacol 9:331
40. Clark LB, Rosen RB, Hartman TG, Louis JB, Suffet I, Lippencott RL, Rosen JD (1992) Intern J Environ Anal Chem 47:167
41. Rudel RA, Melly SJ, Geno PW, Sun G, Brody JG (1998) Environ Sci Technol 32:861
42. Sumpter JP (1995) Toxicol Lett 82/83:737
43. Talmage SS (1994) Environmental and human safety of major surfactants. Lewis Publishers, Tokyo, p 191
44. Talmage SS (1994) Environmental and human safety of major surfactants. Lewis Publishers, Tokyo p 252
45. White R, Jobling S, Hoare SA, Sumpter JP, Parker MG (1994) Endocrinology 135:175
46. APE Research Council (1999) White paper on alkylphenols and alkylphenol ethoxylates
47. Talmage SS (1994) Environmental and human safety of major surfactants. Lewis Publishers, Tokyo, p 255
48. Ahel M, Giger W, Koch M (1994) Water Res 28:1131
49. Bennie DT, Sullivan CA, Lee H-B (1998) Water Qual Res J Can 33:231
50. Naylor CG, Mieure JP, Adams WJ, Weeks JA, Castaldi FJ, Ogle LD, Romano RR (1992) J Amer Oil Chem Soc 69:695
51. Swisher RD (1987) Surfactant biodegradation, 2nd edn. Marcel Dekker, New York, p 47
52. Marcomini A, Capel PD, Lichtensteiger T, Brunner PH, Giger W (1989) J Environ Qual 18:523
53. Reinhard M, Goodman NL, Barker JF (1984) Environ Sci Technol 18:953
54. Ahel M, Schaffner C, Giger W (1996) Water Res 30:37
55. Talmage SS (1994) Environmental and human safety of major surfactants. Lewis Publishers, Tokyo, p 239
56. Ahel M, McEvoy J, Giger W (1993) Environ Pollut 79:243
57. Lewis SK, Lech JJ (1996) Xenobiotica 26:813
58. Coldham NG, Sivapathasundaram S, Dave M, Ashfield LA, Pottinger TG, Goodall C, Sauer MJ (1998) Drug Metab Dispos 26:347
59. Shiraishi H, Carter DS, Hites RA (1989) Biomed Environ Mass Spectrom 18:478
60. Meyer O, Andersen PH, Hansen EV, Larsen JC (1988) Pharmacol Toxicol 62:236
61. Purdom CE, Hardiman PA, Bye VJ, Eno NC, Tyler CR, Sumpter JP (1994) Chem Ecol 8:275
62. Jobling S, Sheahan D, Osborne JA, Matthiessen P, Sumpter JP (1996) Environ Toxicol Chem 15:194
63. Arukwe A, Forlin L, Gorsoyr A (1997) Environ Toxicol Chem 16:2576
64. Gray MA, Metcalf CD (1997) Environ Toxicol Chem 16:1082
65. Silverberg E, Lubera JA (1989) CA Cancer J Clin 39:3
66. Saunders PTK, Majdic G, Parte P, Millar MR, Fisher JS, Turner KJ, Sharpe RM (1997) Fetal and perinatal influence of xenoestrogens on testis gene expression. In: Ivell R, Holstein A-F (eds) The fate of the male germ cell. Plenum Press, New York, p 99
67. Lee PC (1998) Endocrine 9:105
68. Sharpe RM, Fisher JS, Millar MM, Jobling S, Sumpter JP (1995) Environ Health Perspect 103:1136
69. Sharpe RM, Turner KJ, Sumpter JP (1998) Environ Health Perspect 106:220 (Abstract)
70. Smith WF (1994) Fundamentos de la Ciencia e Ingenieria de Materiales. 2nd edn. McGraw-Hill, New York. Cited in: Perez P, Pulgar R, Olea-Serrano F, Villalobos M, Rivas A, Metzler M, Pedraza V, Olea N (1998) Environ Health Perspect 106:167
71. Krishnan AV, Starhis P, Permuth SF, Tokes L, Feldman D (1993) Endocrinology 132:2279
72. Dodds EC, Lawson W (1936) Nature 137:996

73. Brotons JA, Olea-Serrano MF, Villalobos M, Olea N (1994) Environ Health Perspect 103:608
74. Soto AM, Sonnenschein C, Chung KL, Fernandez MF, Olea N, Olea-Serrano MF (1995) Environ Health Perspect 103:113
75. Steinmetz R, Brown NG, Allen DL, Bigsby RM, Ben-Jonathan N (1997) Endocrinology 138:1780
76. Gaido KW, Leonard LS, Lovell S, Gould JC, Babai D, Portier CJ (1997) Toxicol Appl Pharmacol 43:205
77. Kuiper GG, Lemmen JG, Carlsson B, Corton JC, Safe SH, van der Saag PT, van der Burg B, Gustafsson JA (1998) Endocrinology 139:4252
78. Kloas W, Lutz I, Einspanier R (1999) Sci Total Environ 225:59
79. Fang H, Tong W, Perkins R, Soto AM, Prechtl NV, Sheehan DM (2000) Environ Health Perspect (submitted)
80. Soto AM, Chung KL, Sonnenschein C (1994) Environ Health Perspect 102:380
81. Hammond GL (1995) Trends Endocrinol Metab 6:298
82. Nagel SC, vom Saal FS, Thayer KA, Dhar MG, Boechler M, Welshons WV (1997) Environ Health Perspect 105:70
83. Damassa DA, Lin TM, Sonnenschein C, Soto AM (1991) Endocrinology 129:75
84. Dohler KD, Jarzab B (1992) The influence of hormones and hormone antagonists on sexual differentiation of the brain. In: Colborn T, Clement C (eds) Chemically-induced alteration in sexual and functional development: the wildlife/human connection. Princeton Scientific Publishing, Princeton, p 231
85. Milligan SR, Khan O, Nash M (1998) Gener Comp Endocrinol 112:89
86. Staples CA, Dorn PB, Klecka GM, O'Block ST (1998) Chemosphere 36:2149
87. Sheftel VO (1995) Handbook of toxic properties of monomers and additives. CRC Press, Boca Raton
88. Paseiro-Losada P, Simal-Lozano J, Paz-Abuin S, Lopez-Mahia P, Simal-Gandara J (1993) J Anal Chem 345:527
89. Biles JE, McNeal TP, Begley TH, Hollifield HC (1997) J Agric Food Chem 45:3541
90. Lambert C, Larroque M (1997) J Chromat Sci 35:57
91. Howe SR, Borodinsky L, Lyon RS (1998) J Coat Tech 70:69
92. Soderholm K-J, Mariotti A (1999) J Am Dent Assoc 130:201
93. Olea N, Pulgar R, Perez P, Olea-Serrano F, Rivas A, Novillo-Fertrell A, Pedraza V, Soto AM, Sonnenschein C (1996) Environ Health Perspect 104:298
94. Hamid A, Hume WR (1997) Dent Mater 13:98
95. Nathanson D, Lertpitayakun P, Lamkin MS, Edalatpour M, Chou LL (1997) J Am Dent Assoc 128:1517
96. Mariotti A, Soderholm KJ, Johnson S (1998) Eur J Oral Sci 106:1022
97. Raveh J, Stich H, Schawalder C, Ruchti D, Cottier H (1982) Acta Otolaryng 94:371
98. Vuillemin T, Raveh J, Stich H, Cottier H (1987) Arch Oto Head Neck Surg 116:836
99. Tamura J, Kawanabe K, Kobayashi M, Nakamura T, Kokubo T, Yoshihara S, Shibuya T (1996) J Biomed Mat Res 30:85
100. Kobayashi M, Nakamura T, Tamura J, Kokubo T, Kikutani T (1997) J Biomed Mat Res 37:301
101. Tamura J, Kitsugi T, Iida H, Fugita H, Nakamura T, Kokubo T, Yoshihara S (1995) Bioceramics 8:219
102. Matsuda Y, Ido K, Nakamura T, Fujita H, Yamamuro T, Oka M, Shibuya T (1997) Clin Ortho Rel Res 336:263
103. Climie IJG, Hutson DH, Stoydin G (1981) Xenobiotica 11:391
104. Allen H, Kaidbey K (1979) Arch Derm 115:1307
105. Knaak JB, Sullivan LJ (1966) Toxicol Appl Pharmacol 8:175
106. Takahashi O, Oishi S (2000) Environ Health Perspect 108:931
107. Atkinson A, Roy D (1995) Biochem Biophys Res Comm 210:424
108. Atkinson A, Roy D (1995) Environ Molec Mutag 26:60
109. Climie IJG, Hutson DH, Stoydin G (1981) Xenobiotica 11:401

110. Stone CM, Watkinson RJ (1983) Sittingbourne Research Centre, Rep SBGR 83:425, Kent, England
111. Spivack J, Leib TK, Lobos JH (1994) J Biol Chem 269:7323
112. Hardin BD, Bond GP, Sikov MR, Andrew FD, Beliles RP, Niemeier RW (1981) Scand J Work Environ Health 7 Suppl 4:66
113. Morrissey RE, George JD, Price CJ, Tyl RW, Marr MC, Kimmel CA (1987) Fund Appl Toxicol 8:571
114. Nonneman DJ, Ganjam VK, Welshons WV, vom Saal FS (1992) Biol Reprod 47:723
115. vom Saal FS, Timms BG, Montano MM, Palanza P, Thayer KA, Nagel SC, Dhar MD, Ganjam VK, Parmigiani S, Welshons WV (1997) Proc Natl Acad Sci USA 94:2056
116. vom Saal FS, Cooke PS, Buchanan DL, Palanza P, Thayer KA, Nagel SC, Parmigiani S, Welshons WV (1998) Toxicol Ind Health 14:239
117. Sheehan DM, Willingham E, Gaylor D, Bergeron JM, Crews D (1999) Environ Health Perspect 107:155
118. Howdeshell KL, Hotchkiss AK, Thayer KA, Vandenbergh JG, vom Saal FS (1999) Nature 401:763
119. Rubin BS, Murray MK, Damassa DA, King JC, Soto AM (2001) Environ Health Perspect 109:675
120. Sonnenschein C, Olea N, Pasanen ME, Soto AM (1989) Cancer Res 49:3474
121. Soto AM, Lin TM, Sakabe K, Olea N, Damassa DA, Sonnenschein C (1995) Oncology Res 7:545
122. Markey CM, Luque EH, Munoz de Toro M, Sonnenschein C, Soto AM (in press) Biol Reprod
123. Evans JS, Varney RF, Koch FC (1941) Endocrinology 28:747
124. Rubin BL, Dorfman AS, Black L, Dorfman RI (1951) Endocrinology 49:429
125. Coldham NG, Dave M, Sivapathasundaram S, McDonnell DP, Connor C (1997) Environ Health Perspect 105:734
126. Markey CM, Michaelson CL, Sonnenschein C, Soto AM (2001) Environ Health Perspect 109:55
127. Ashby J, Tinwell H (1998) Environ Health Perspect 106:719
128. Dodge JA, Glasebrook AL, Magee DE, Phillips DL, Sato M, Short LL, Bryant HU (1996) J Steroid Biochem Molec Biol 59:155
129. Steinmetz R, Mitchner NA, Grant A, Allen DL, Bigsby RM, Ben-Jonathan N (1998) Endocrinology 139:2741
130. Colerangle JB, Roy D (1997) J Steroid Biochem Molec Biol 60:153
131. Lamb J (1997) Environ Health Perspect 105:273
132. Liaw JJ, Gould JC, Welsch F, Sar M (1997) Toxicologist 36:14 (Abstract No 72)

CHAPTER 8

Hydroxylated Polychlorinated Biphenyls (PCBs) and Organochlorine Pesticides as Potential Endocrine Disruptors

Stephen Safe

Department of Veterinary Physiology and Pharmacology, Texas A & M University, College Station, TX 77843-4466, USA
e-mail: ssafe@cvm.tamu.edu

Hydroxylated polychlorinated biphenyls (hydroxy-PCBs) and various organochlorine pesticides have been identified as environmental contaminants, and there has been some concern regarding their potential adverse effects as endocrine-active agents. Initial studies showed that two synthetic compounds, 2′,4′,6′-trichloro- and 2′,3′,4′,5′-tetrachloro-4-biphenylol (HO-PCB3 and HO-PCB4, respectively), bound to the estrogen receptor (ER) and exhibited estrogenic activity in both *in vivo* and *in vitro* assays. Although these compounds alone were weakly estrogenic (>1000 times less potent than 17β-estradiol), there were some reports suggesting that interactions of these compounds and several organochlorine pesticides were synergistic. Subsequent studies in this laboratory have confirmed that HO-PCB3 and HO-PCB4 were weakly estrogenic and equimolar concentrations of these compounds gave additive responses. Similar results were obtained for organochlorine pesticides. Structure-activity relationships for hydroxy-PCBs showed that most compounds were either weakly estrogenic or inactive; moreover, several compounds exhibited antiestrogenic activity. Based on their weak estrogenic potencies and low environmental levels, it is unlikely that hydroxy-PCB and organochlorine pesticides contribute significantly to overall xenoestrogen action.

Keywords. Hydroxy-PCBs, Estrogenicity, Organochlorine pesticides

1 **Introduction** 156

2 **Hydroxy-PCBs: Origins and Environmental Occurrence** 156

3 **Hydroxy-PCBs as Estrogens and Antiestrogens** 157

3.1 HO-PCB3 and HO-PCB4 Interactions 157
3.2 HO-PCBs as Estrogens/Antiestrogens – Structure-Activity Relationships 159
3.3 Hydroxy-PCBs in Humans – Estrogenic/Antiestrogenic Activities 162

4 **Hydroxy-PCBs and Thyroid Function** 163

5 **Organochlorine Pesticides as Endocrine Disruptors** 163

6 **Conclusions** 164

7 **References** 165

The Handbook of Environmental Chemistry Vol. 3, Part L
Endocrine Disruptors, Part I
(ed. by M. Metzler)

Abbreviations

CAT	chloramphenicol acetyltransferase
CKB	creatine kinase B
p,p′-DDE	2,2-bis(*p*-chlorophenyl)-1,1-dichloroethylene
o,p′-DDT	2-(*p*-chlorophenyl)-2-(*o*-chlorophenyl)-1,1,1-trichloroethane
p,p′-DDT	2,2-bis(*p*-chlorophenyl)-1,1,1-trichloroethane
E2	17*β*-estradiol
ER	estrogen receptor
HO-PCB	hydroxy-polychlorinated biphenyl
luc	luciferase
PR	progesterone receptor
RBA	relative binding affinity

1 Introduction

Recently there has been considerable scientific, regulatory and public concern regarding the potential adverse effects of industrial chemicals and their by-products that act via disruption of endogenous endocrine pathways [1–3]. The hypotheses suggesting that background levels of organochlorine environmental contaminants such as 2,2-bis(*p*-chlorophenyl)-1,1-dichloroethylene (*p,p′*-DDE), polychlorinated biphenyls (PCBs), dibenzofurans and dibenzo-*p*-dioxins and related compounds may be causally related to decreased male reproductive capacity, an increased incidence of breast cancer and neurodevelopmental deficits in children are based on environmental, laboratory animal and human data. In 1992, a report by Carlsen and coworkers showed that meta-analysis of 61 sperm count studies from 1940 to 1990 indicated the sperm counts worldwide had decreased from 113×10^6 to 66×10^6/mL [4]. These data coupled with increasing incidence of testicular cancer in most countries led to the hypothesis that *in utero* exposure to environmental estrogens and possibly antiandrogens such as *p,p′*-DDE [5] were causally linked to decreased male reproductive capacity. This hypothesis was biologically plausible since it is well known that offspring from women or laboratory animals exposed to the potent estrogenic drug diethylstilbestrol exhibit a host of male and female reproductive tract abnormalities [6]. These problems have also been observed in some wildlife populations exposed to other endocrine-active contaminants [1, 2, 7]. This chapter will briefly review the endocrine-like activity of two important classes of environmental contaminants, namely hydroxy-PCBs and organochlorine pesticides.

2 Hydroxy-PCBs: Origins and Environmental Occurrence

PCB mixtures are industrial compounds that have been widely identified as persistent environmental contaminants that bioaccumulate in fish, wildlife and humans. In laboratory animal studies, PCBs induce a diverse spectrum of biochemical and toxic responses that are due, in part, to disruption of endocrine pathways [8–10]. In addition, PCBs are metabolized to give multiple products including hydroxylated metabolites and their conjugates, dihydrodiols, catechols and several methylsulfonyl analogues [11]. Hutzinger and coworkers first showed that diverse species metabolized individual PCB congeners and mixtures to give hydroxylated metabolites that are readily excreted in urine and feces [12]. Cytochrome P450-mediated oxidative metabolism of PCBs has been well characterized and is an important pathway for detoxification and removal of the parent hydrocarbons [11, 13]. Results of a few studies demonstrated that hydroxy-PCB metabolites were less toxic than their parent hydrocarbons [14, 15]. Surprisingly, Bergman and coworkers demonstrated that hydroxy-PCBs were relatively persistent in serum [16], and they identified a diverse spectrum of highly chlorinated hydroxy-PCBs in human serum, wildlife samples, and in laboratory rats administered commercial PCB mixtures[17]. Subsequent studies have confirmed that hydroxy-PCBs persist in serum of humans and other species [18–20] and therefore effects of these compounds on endocrine response pathways may be adverse.

3 Hydroxy-PCBs as Estrogens and Antiestrogens

Korach and coworkers [21] first showed that hydroxy-PCBs competitively bound the mouse estrogen receptor (ER) and two of the most active ER agonists in their study were 2′,4′,6′-trichloro-4-biphenylol (HO-PCB3) and 2′,3′,4′,5′-tetrachloro-4-biphenylol (HO-PCB4). Structure-activity relationships suggested that the most active congeners contained a single *para*-hydroxy group on one of the two biphenyl rings. Interest in hydroxy-PCBs as endocrine disruptors was heightened by two studies using HO-PCB3 and HO-PCB4 as model compounds. Bergeron and coworkers [22] utilized turtles as a model system for investigating the estrogenic activity of HO-PCB3 and HO-PCB4. Turtle eggs incubated at 26 °C result in 100% males whereas at 32 °C all females are produced. However, temperature-dependent sex reversal can also obtained with 17β-estradiol (E2) and they also reported that incubation of turtle eggs with HO-PCB3 and HO-PCB4 resulted in dose-dependent sex reversal (males → females). Moreover, it appeared that a combination of both hydroxy-PCBs synergistically induced sex reversal. The issue of synergistic interaction of hydroxy-PCBs was also addressed in a paper by Arnold and coworkers [23] that reported synergistic interactions of HO-PCB3 and HOPCB4 in receptor binding and transactivation assays. This paper was subsequently withdrawn [24]; nevertheless, non-additive interactions of weakly estrogenic pesticides have been suggested by other investigators.

3.1 HO-PCB3 and HO-PCB4 Interactions

Studies were initiated in this laboratory [25] to investigate estrogenic activities of HO-PCB3, HO-PCB4 and their combination in several E2-responsive assays including: binding to the mouse uterine ER; induction of immature mouse uter-

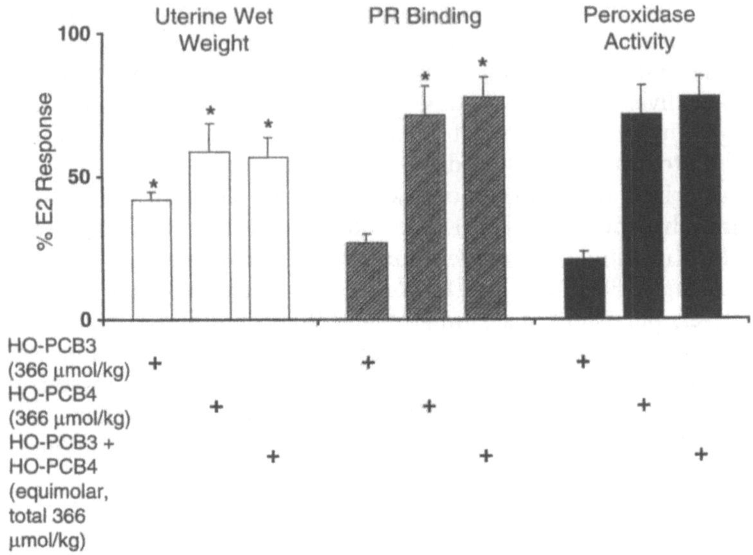

Fig. 1. Estrogenic activity of HO-PCB3, HO-PCB4 and their combination (equimolar) in 21-day-old B6C3FI mice. Mice were administered the test compounds alone or in combination on 3 consecutive days; uterine wet weight, peroxidase activity and PR binding were determined [25] and compared to results obtained with E2 alone (0.0053 μmol/kg for three consecutive days)

ine ER and PR binding, peroxidase activity and wet weight; induction of MCF-7 human breast cancer cell growth; induction of luciferase activity in human HepG2, MDA-MB-231 and MCF-7 cells transiently transfected with pC3-luc (complement C3 gene promoter linked to luciferase reporter gene); induction of chloramphicol acetyltransferase (CAT) activity in MCF-7 cells transiently transfected with pCKB (rat creatine kinase B promoter linked to a CAT reporter gene). In addition, an E2-responsive yeast-based assay was also used. The results clearly demonstrated that both HO-PCB3 and HO-PCB4 were nearly full ER agonists in all *in vitro* assays; however, in the mouse uterine assay (Fig. 1) submaximal induction was observed at doses as high as 366 μmol/kg/day (for 3 days) and both hydroxy-PCB congeners were >70,000 times less potent than E2 (Fig. 1). In most of the *in vivo* and *in vitro* bioassays, both HO-PCB3 and HO-PCB4 (10^{-5} M) induced >50% of the maximal response observed for 10^{-9} M E2 (Fig. 2) and were generally >10,000 times less active than E2. The major exception was in the MCF-7 cell proliferation and yeast-based assays, where differences in estrogenic potency were only 2 to 3 orders of magnitude. Interactions of HO-PCB3 and HO-PCB4 were examined using equimolar mixtures and the results were additive for most responses as illustrated in Figs. 1 and 2. The apparent small differences from non-additivity observed in the *in vivo* studies (Fig. 1) may be related to differences in animal-responsiveness since HO-PCB3 was investigated in one study and results for HO-PCB4 and the equimolar mixture were determined in a second study, and the E2-induced progesterone re-

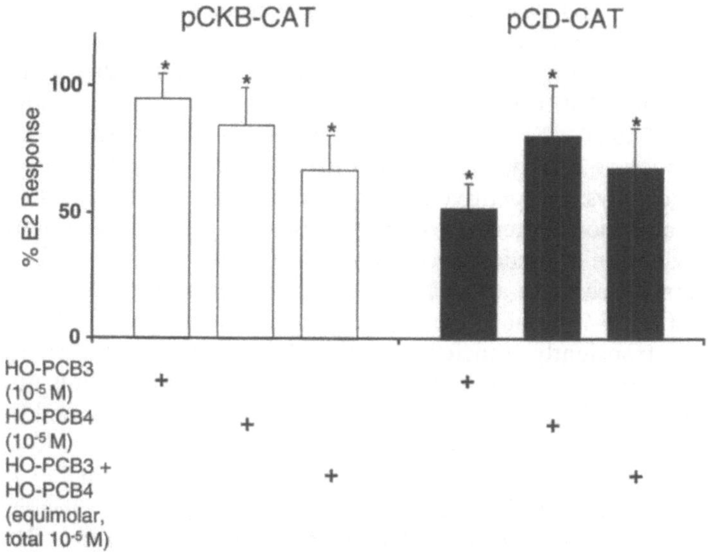

Fig. 2. Estrogenic activity of HO-PCB3, HO-PCB4 and their equimolar combination in MCF-7 cells transfected with construct derived from the E2-responsive creatine kinase B (pCKB-CAT) and cathepsin D (pCD-CAT) gene promoters. The results obtained for HO-PCBs are compared to responses observed for E2 alone (10^{-8} M) [25]

ceptor (PR) binding and peroxidase activity were significantly higher in the first study.

There was some initial concern that non-additive interactions of HO-PCBs and other weakly estrogenic compounds would be observed only at low levels of ER expression. This point was addressed by determining the effects of HO-PCB3, HO-PCB4 and their combination in HepG2 and MDA-MB-231 cells transiently transfected with pC3-luc and variable amounts of ER expression plasmid. The results showed that induction responses decreased with decreasing ER levels; however, the effects of HO-PCB3/HO-PCB4 mixture (equimolar) were similar to those observed for the compounds alone. These results clearly showed that for an equimolar binary mixture of HO-PCB3 and HO-PCB4, their estrogenic activities were additive in a battery of *in vitro* and *in vivo* assays.

3.2
HO-PCBs as Estrogens/Antiestrogens – Structure-Activity Relationships

Korach and coworkers [21] first reported the estrogenic activity of hydroxy-PCBs, and HO-PCB3 and HO-PCB4 were the most potent congeners among the 12 different mono- and dihydroxy-substituted biphenyls. Structure-activity relationships among these compounds were not apparent due to their structural diversity and therefore a more systematic structure-estrogenicity/antiestrogenicity study for HO-PCBs was initiated. Optimal estrogenic activity was previously observed for congeners containing a single *p*-hydroxy group on one phenyl ring (e.g., HO-PCB3 and HO-PCB4), therefore the effects of additional chlorine substitution *ortho* or *meta* to the hydroxy group were determined using a series of 3-chloro-4-hydroxyphenyl and 2-chloro-4-hydroxyphenyl analogues [26]. The substitution patterns on the chlorinated ring included 2′,4′,6′-trichloro-, 2′,3′,5′,6′-, 2′,3′,4′,6′- and 2′,3′,4′5′-tetrachlorophenyl (Fig. 3). Subsequent structure-estrogenicity relationships were determined using the following bioassays: competitive binding to the rat and mouse cytosolic ER; immature rat and mouse uterine wet weight, uterine peroxidase activity and PR binding; induction of luciferase activity in HeLa cells stably transfected with a Gal4: human ER chimera and a 17mer-regulated luciferase reporter gene; proliferation of MCF-7 human breast cancer cells; induction of CAT activity in MCF-7 cells transiently transfected with a full-length human ER expression plasmid and a plasmid containing an estrogen-responsive vitellogenin A2 gene promoter insert linked to a CAT reporter gene. The results obtained from these assays were highly variable and assay-dependent. For example, an IC_{50} value for competitive binding to the rat or mouse ER was not observed for 2′,3,4′,6′-tetrachloro-4-biphenylol, whereas the remaining compounds gave variable IC_{50} values that were not structure-dependent. Relative binding affinities (RBAs) compared to E2 (RBA = 1.0) varied from 5.3×10^{-6} to 1.4×10^{-3} and 1.3×10^{-4} to 7.2×10^{-4} for cytosolic uterine rat and mouse ER, respectively. Dose-dependent induction of uterine wet weight, peroxidase activity and PR binding was not observed for any of the hydroxy-PCB congeners 1–8 in the immature mouse or rat uterus at doses as high as 100 mg/kg/day (for 3 days). Structure-estrogenicity relationships for HO-PCBs 1–8 were not observed in MCF-7 cells transfected with pVit-A2 since only 2 compounds, 2,2′,3′,4′,6′- and 2,2′,3′,4′,5′-pentachloro-4-biphenylol significantly induced CAT activity at the highest concentration (10^{-5} M). In contrast, comparable structure-estrogenicity relationships were observed for MCF-7 cell proliferation assays and induction of luciferase activity in stably transfected HeLa cells. Four compounds, namely 2,2′,3′,4′,6- and 2′,3,3′,4′,6′-pentachloro-, 2,2′,4′,6′- and 2′,3,4,6′-tetrachloro-4-biphenylol, exhibit ER agonist activity for both responses. In contrast, 2,2′,3′,4′,5′-, 2′,3,3′, 4,5′-, 2,2′,3′,5′,6′- and 2′,3,3′,5′,6′-pentachloro-4-biphenylol did not induced estrogenic responses (i.e., see Fig. 4). These results demonstrate that chloro substitution in the phenolic ring did not significantly affect activity, whereas 2,4,6-trichloro- and 2,3,4,6-tetrachloro-substitution on the chlorophenyl ring gave compounds with the highest estrogenic activity.

Since the hydroxy-PCB congeners (Fig. 3) were weak ER agonists, we also investigated their antiestrogenic activities in the same assays. Structure-antiestrogenicity assays were inconsistent in the rodent uterus and in MCF-7 cells transfected with pVit-CAT. Several hydroxy-PCBs inhibited one or more E2-induced responses in the mouse uterus and one compound, 2,2′,3′,4′,6′-pentachloro-4-biphenylol, significantly inhibited all three responses (uterine wet weight increase, PR binding, and peroxidase activity). In contrast, this compound was not a consistent ER antagonist in the *in vitro* assays, and none of the hydroxy-PCBs inhibited E2-induced responses in the immature rat uterus. Interestingly, structure-antiestrogenicity was observed for inhibition of E2-induced cell proliferation and luciferase activity in stabily-transfected HeLa cells;

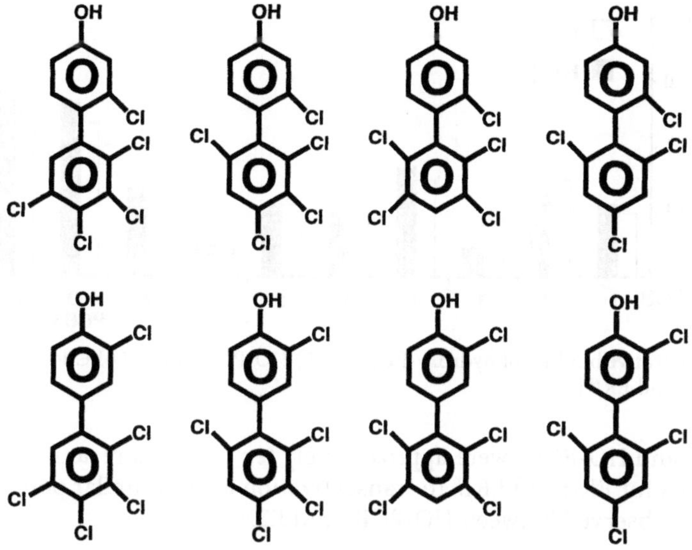

Fig. 3. Hydroxy-PCB congeners used in structure-activity studies [26]

the 2′,3′,4′,5′- and 2′,3′,5′,6′-tetrachlorophenyl-substituted compounds exhibited antiestrogenic activity whereas the weakly estrogenic 2′,4′,6′-trichloro- and 2′,3′,4′,6′-tetrachloro-substituted analogues were not antiestrogenic in these assays (Fig. 4). Thus, patterns of estrogenic and antiestrogenic activity for this series of hydroxy-PCBs were highly response-dependent and not readily predicted from structure.

Kuiper and coworkers [27] recently investigated the RBAs and transcriptional activation of HO-PCB3, HO-PCB4 and the 8 congeners illustrated in Fig. 3 using human ERα and ERβ. HO-PCBs 1–8 bound to both ERα and ERβ with comparable affinities but were 200 to 3300 times less active (Fig. 5). In contrast,

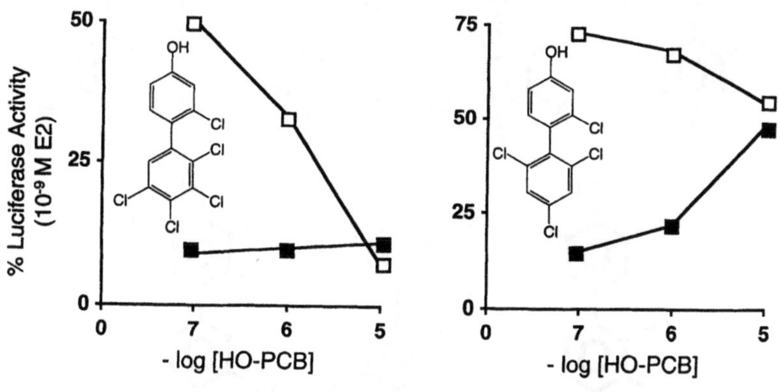

Fig. 4. Estrogenic and antiestrogenic activities of two hydroxy-PCB congeners in stabily-transfected HeLa cells. Cells were treated with different concentrations of hydroxy-PCBs alone (■) or in combination with 6×10^{-11} M E2 (□) as described [26]

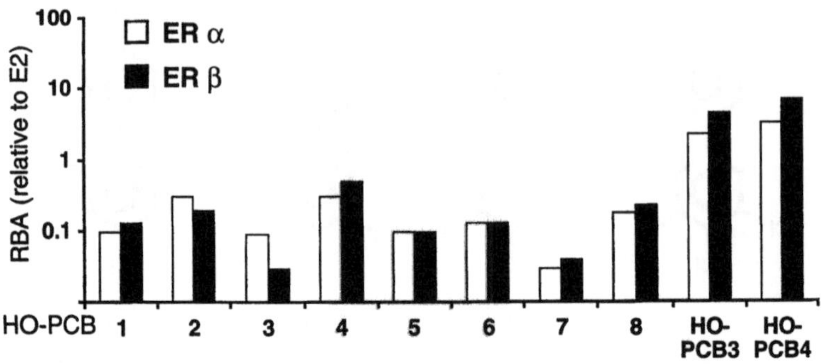

Fig. 5. Comparative binding of hydroxy-PCBs (1–8) HO-PCB3 and HO-PCB4 to recombinant human ERα and ERβ [27]

HO-PCB4 and HO-PCB3 were approximately ten times more active as competitive ligands for ERα and ERβ. In transactivation assays, similar potency differences were observed between HO-PCBs and E2.

3.3 Hydroxy-PCBs in Humans – Estrogenic/Antiestrogenic Activities

Identification of serum-persistent hydroxy-PCBs in humans (Fig. 6) stimulated research on the estrogenic and antiestrogenic activities of these congeners [28, 29]. Competitive binding to rat cytosolic ER, human ERα and human ERβ was minimal for all these compounds and this paralleled their effects in transactivation assays [27–29]. In contrast, all of the hydroxy-PCBs identified in humans inhibited one or more E2-induced responses and one congener, 2,2′,3,4′,5,5′,6-

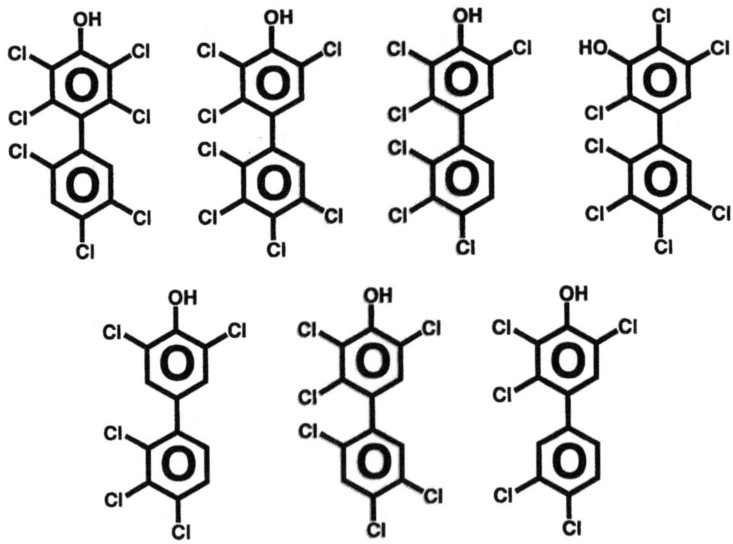

Fig. 6. Hydroxy-PCBs identified in human serum

heptachloro-4-biphenylol, inhibited E2-induced transactivation (in HeLa and MCF-7 cells) and proliferation of MCF-7 cells. These results suggest that the contribution of hydroxy-PCBs to overall human exposure to xenoestrogens is minimal.

4 Hydroxy-PCBs and Thyroid Function

Several studies have demonstrated that PCBs and other halogenated aromatic hydrocarbons alter thyroid hormone status in rodents; this has been attributed, in part, to altered conjugation of triiodothyronine or thyroxine by induced phase II drug metabolizing enzyme activities [30–33]. Brouwer and coworkers initially discovered that unexpectedly high serum levels of hydroxy-PCBs and other halogenated aromatic hydrocarbon metabolites were due to their competitive binding to transthyretin, the major thyroid hormone serum transport protein in rodents [34–36]. It has been suggested that interaction of hydroxy-PCBs with transthyretin can lower circulating thyroxine levels and affect fetal thyroid hormone uptake. This pathway may be important for subsequent fetal development and is an area of concern regarding adverse impacts of hydroxylated aromatics on various wildlife species. For example, exposure of pregnant rats to 3,3′,4 4′-tetrachlorobiphenyl gives relatively high concentrations of the 3,3′,4′,5-tetrachloro-4-biphenylol metabolite that binds with high affinity to transthyretin and accumulates in fetal brains [37]. Moreover, similar results were obtained with the PCB mixture Aroclor 1254 in which selective hydroxy-PCB metabolites can also be detected in fetal brain [38]. The role of these compounds in fetal development and in the offspring may be important in some exposure scenarios; however, their role in human health is unknown.

5 Organochlorine Pesticides as Endocrine Disruptors

Organochlorine pesticides such as 2,2-bis(*p*-chlorophenyl)-1,1,1-trichloroethane (*p,p*′-DDT) and related compounds have been extensively used for pest control; however, detection of these compounds in fish, wildlife and humans has resulted in restricted use or banning of many of these bioaccumulative compounds. Nevertheless, many organochlorine pesticides or their metabolites have been detected in environmental samples or food products and *p,p*′-DDE is a common persistent pollutant identified in most biotic samples including humans [39, 40]. Some of the earliest studies on the endocrine activity of organochlorine pesticides utilized the rodent uterus model to study estrogenic/antiestrogenic activities and most of the data indicated minimal effects and these included some antiestrogenic responses that may have been related to enhanced cytochrome P450 levels and increased estrogen metabolism [41]. In addition, several studies clearly demonstrated that kepone and 2-(*p*-chlorophenyl)-2-(*o*-chlorophenyl)-1,1,1-trichloroethane (*o,p*′-DDT) competitively bound the ER and exhibited ER agonist activity *(in vivo/in vitro)* [42–44]; however, these compounds are not routinely detected as widespread environmental contaminants. More recently, Soto and coworkers showed that several additional pesticides stimulated MCF-7 human breast cancer cell growth in the E-screen assay, and these include endosulfan, toxaphene and dieldrin; these compounds were approximately 10,000 times less active than E2 [45, 46].

Major concern regarding estrogenic activities of organochlorine pesticides followed the report by Arnold and coworkers [23] showing that interactions of weak environmental estrogens dieldrin, endosulfan, toxaphene and chlordane gave synergistic ER binding and transactivation responses. For example, "combinations of two weak environmental estrogens such as dieldrin, endosulfan or toxaphene, were 1000 times more potent in hER-mediated transactivation than any chemical alone". These synergistic interactions could not be repeated in other laboratories [47–49], and the original paper was subsequently withdrawn [24].

Although the concern regarding the estrogenic activity of organochlorine pesticides has decreased, Kelce and coworkers [5] reported that *p,p′*-DDE bound the androgen receptor (AR) and exhibited a broad spectrum of antiandrogenic activities in both *in vivo* and *in vitro* bioassays. For example, *in utero* exposure of pregnant rats to *p,p′*-DDE resulted in decreased anogenital distance in male pups at birth and retained thoracic nipples on postnatal day 13. *p,p′*-DDE-induced antiandrogenic activity was also observed in immature and adult male rats. Many of these same responses were also observed in another study utilizing Sprague-Dawley and Long Evans Hooded rats [50] and their results also showed that effects on male rat sexual differentiation were minimal at maternal doses below 10 mg/kg (administered daily from gestation day 14 to 18). Adipose tissue levels in rat pups administered 10 or 100 mg/kg varied from 0.77 to 7.27 ppm, respectively, and *p,p*-DDE levels of >1 ppm were commonly observed in human populations in the 1970s and 1980s prior to restrictions on applications of *p,p′*-DDT. Fortunately, environmental and human levels of *p,p′*-DDE have declined rapidly in most locations; however, the adverse impacts of *p,p′*-DDE and related compounds as antiandrogens requires further study.

6
Conclusions

It is clear that both hydroxy-PCBs and organochlorine pesticides modulate endocrine responses in animal models and cells in culture. However, there is also considerable debate regarding a linkage between background levels of endocrine-active chemicals and decreased male reproductive capacity and these are briefly summarized below.

1. Recent studies show both decreased or unchanged sperm counts in subjects from diverse clinics; moreover, important demographic differences have been observed in the United States, France, Denmark, Scandinavia and Canada [51–58]. Demography was not taken into account in the original meta-analysis by Carlsen and coworkers [4].
2. It is unlikely that organochlorine environmental contaminant levels are variable within regions that show major sperm count variability [39].

3. Although the incidence of testicular cancer is higher in Denmark vs. Finland, *p,p'*-DDE levels in both countries are similar and have been decreasing (80–90%) over the past 30+ years [59, 60].
4. Correlations between decreased male reproductive capacity and any environmental contaminant have not been reported.

The hypothesis that environmental estrogens (or xenoestrogens) are preventable causes of breast cancer [61] was primarily supported by two small case-control studies showing that serum DDE or tissue PCB levels were higher in breast cancer patients vs. controls [62, 63]. Subsequent studies in Europe, United States and Mexico with a large number of cases/controls have shown that PCB/DDE levels were not elevated in breast cancer patients vs. controls [64–67]. A recent report by the National Research Council concluded that "An evaluation of the available studies conducted to date does not support an association between adult exposure to DDT, DDE, TCDD and PCBs and cancer of the breast" [68] Thus, although results of some human studies do not support the endocrine disruptor hypothesis, further research in the area will help resolve this contentious issue.

Acknowledgement. The financial assistance of the National Institutes of Health (ES04917 and ES09106) and the Texas Agricultural Experiment Station is gratefully acknowledged.

7 References

1. Colborn T, Vom Saal FS, Soto AM (1993) Environ Health Perspect 101:378
2. Thomas KB, Colborn T (1992) Organochlorine endocrine disruptors in human tissue. In: Colborn T, Clement C (eds), Chemically Induced Alterations in Sexual Development: the Wildlife/Human Connection. Princeton Scientific, Princeton, NJ, p 365
3. Sharpe RM, Skakkebaek NF (1993) Lancet 341:1392
4. Carlsen E, Giwercman A, Keiding N, Skakkebaek NE (1992) Br Med J 305:609
5. Kelce WR, Stone CR, Laws SC, Gray LE (1995) Nature 375:581
6. Newbold R (1995) Environ Health Perspect 103:83
7. Giesy JP, Ludwig JP, Tillitt DE (1994) Environ Sci Technol 28:128A
8. Safe S (1984) CRC Crit Rev Toxicol 12:319
9. Safe S (1990) CRC Crit Rev Toxicol 21:51
10. Safe S (1994) CRC Crit Rev Toxicol 24:87
11. Safe S (1989) Polyhalogenated aromatics: uptake, disposition and metabolism. In: Kimbrough RD, Jensen AA (eds), Halogenated Biphenyls, Naphthalene, Dibenzodioxins and Related Compounds, 2nd edn. Elsevier-North Holland, Amsterdam, p 51
12. Hutzinger O, Nash DM, Safe S, Norstrom RJ, Wildesh DJ, Zitko V (1972) Science 178:312
13. Sipes IG, Schnellmann RG (1987) Biotransformation of PCBs: metabolic pathways and mechanisms. In: Safe S, Hutzinger O (eds), Polychlorinated Biphenyls (PCBs): Mammalian and Environmental Toxicology. Springer, Berlin, Heidelberg, New York, p 97
14. Yoshimura H, Yonemoto Y, Yamada H, Koga N, Oguri K, Saeki S (1987) Xenobiotica 17:897
15. Stadnicki SS, Allen JR (1979) Bull Environ Contam Toxicol 23:788
16. Bergman Å, Klasson-Wehler E, Kuroki H (1994) Environ Health Perspect 102:464
17. Sjodin A, Tullsten AK, Klasson-Wehler E (1998) Organochlorine Compounds 37:365
18. Newsome WH, Davies D (1996) Chemosphere 33:559
19. Jansson B, Jensen S, Olsson M, Renberg L, Sundström G, Vaz R (1975) Ambio 4:93
20. Klasson-Wehler E, Kuroki H, Athanasiadou M, Bergman Å (1992) Organohalogen Compounds, Dioxin '92 10:121

21. Korach KS, Sarver P, Chae K, McLachlan JA, McKinney JD (1988) Mol Pharmacol 33:120
22. Bergeron JM, Crews D, McLachlan JA (1994) Environ Health Perspect 102:786
23. Arnold SF, Klotz DM, Collins BM, Vonier PM, Guillette LJ Jr, McLachlan JA (1996) Science 272:1489
24. McLachlan JA (1997) Science 277:462
25. Ramamoorthy K, Vyhlidal C, Wang F, Chen I-C, Safe S, McDonnell DP, Leonard LS, Gaido KW (1997) Toxicol Appl Pharmacol 147:93
26. Connor K, Ramamoorthy K, Moore M, Mustain M, Chen I, Safe S, Zacharewski T, Gillesby B, Joyeux A, Belaguer P (1997) Toxicol Appl Pharmacol 145:111
27. Kuiper GG, Lemmen JIG, Carlsson B, Corton JC, Safe SH, Van der Saag PT, Van der Burg B, Gustafsson J-A (1998) Endocrinology 139:4252
28. Kramer VJ, Helferich WG, Bergman Å, Klasson-Wehler E, Giesy JP (1997) Toxicol Appl Pharmacol 144:363
29. Moore M, Mustain M, Daniel K, Safe S, Zacharewski T, Gillesby B, Joyeux A, Balaguer P (1997) Toxicol Appl Pharmacol 142:160
30. Brouwer A, Morse DC, Lans MC, Schuur AG, Murk AJ, Klasson-Wehler E, Bergman Å, Visser TJ (1998) Toxicol Ind Health 14:59
31. Schuur AG, Boekhorst FM, Brouwer A, Visser TJ (1997) Endocrinology 138:3727
32. Schuur AG, Tacken PJ, Visser TJ, Brouwer A (1998) Environ Toxicol Pharmacol 5:7
33. Schuur AG, Legger FF, Van Meeteren ME, Moonen MJH, Van Leeuwen-Bol I, Bergman Å, Visser TJ, Brouwer A (1998) Chem Res Toxicol 11:1075
34. Brouwer A, Van den Berg KJ (1986) Toxicol Appl Pharmacol 85:301
35. Lans MC, Klasson-Wehler E, Willemsen M, Meussen E, Safe S, Brouwer A (1993) Chem Biol Interact 88:7
36. Brouwer A, Klasson-Wehler E, Bokdam M, Morse DC, Traag WA (1990) Chemosphere 20:1257
37. Morse DC, Klasson-Wehler E, van de Pas M, de Bie AT, Van Bladeren PJ, Brouwer A (1995) Chem Biol Interact 95:41
38. Morse DC, Klasson-Wehler E, Wesseling W, Koeman JH, Brouwer A (1996) Toxicol Appl Pharmacol 136:269
39. Kutz FW, Wood PH, Bottimore DP (1991) Rev Environ Contamin Toxicol 120:1
40. Winter CK (1992) Rev Environ Contamin Toxicol 127:23
41. Welch RM, Levin W, Conney AH (1969) Toxicol Appl Pharmacol 14:358
42. Hammond B, Katzenellenbogen BS, Krauthammer N, McConnell J (1979) Proc Natl Acad Sci USA 76:6641
43. Robinson AK, Mukku VT, Spalding DM, Stancel GM (1984) Toxicol Appl Pharmacol 76:537
44. Bitman J, Cecil HC (1970) J Agric Food Chem 18:1108
45. Soto AM, Chung KL, Sonnenschein C (1994) Environ Health Perspect 102:380
46. Soto AM, Sonnenschein C, Chung KL, Fernandez MF, Olea N, Serrano FO (1995) Environ Health Perspect 103 (Suppl 7):113
47. Ramamoorthy K, Wang F, Chen I-C, Norris JD, McDonnell DP, Gaido KW, Bocchinfuso WP, Korach KS, Safe S (1997) Science 275:405
48. Ramamoorthy K, Wang F, Chen I-C, Norris JD, McDonnell DP, Gaido KW, Bocchinfuso WP, Korach KS, Safe S (1997) Endocrinology 138:1520
49. Ashby J, Lefebvre PB, Odum J, Harris CA, Routledge EJ, Sumpter JP (1997) Nature 385:494
50. You L, Casanova M, Archibeque-Engle S, Sar M, Fan LQ, Heck HD (1998) Toxicol Sci 45:162
51. Fisch H, Goluboff ET, Olson JH, Feldshuh J, Broder SJ, Barad DH (1996) Fertil Steril 65:1009
52. Rasmussen PE, Erb K, Westergaard LG (1997) Fertil Steril 68:1059
53. Handelsman DJ (1997) Human Reprod 12:101
54. Auger J, Jouannet P (1997) Human Reprod 12:740
55. Zheng Y, Bonde JPE, Ernst E, Mortensen JT, Egense J (1997) Int J Epidemiol 26:1289
56. Fisch H, Goluboff ET (1996) Fertil Steril 65:1044

57. Younglai EV, Collins JA, Foster WG (1998) Fertil Steril 70:76
58. Becker S, Berhane K (1997) Fertil Steril 67:1103
59. Ekbom A, Wicklund-Glynn A, Adami HO (1996) Nature 347:553
60. Cocco P, Benichou J (1998) Oncology 55:334
61. Davis DL, Bradlow HL, Wolff M, Woodruff T, Hoel DG, Anton-Culver H (1993) Environ Health Perspect 101:372
62. Wolff MS, Toniolo PG, Leel EW, Rivera M, Dubin N (1993) J Natl Cancer Inst 85:648
63. Falck F, Ricci A, Wolff MS, Godbold J, Deckers P (1992) Arch Environ Health 47:143
64. Krieger N, Wolff MS, Hiatt RA, Rivera M, Vogelman J, Orentreich N (1994) J Natl Cancer Inst 86:589
65. Van't Veer P, Lobbezoo IR, Martin-Moreno JM, Guallar F, Gomez-Aracena J, Kardinaal AFM, Kohlmeier L, Martin BC, Strain A Thumm M, Van Zoonen P, Baumann BA, Huttunen JK, Kok FJ (1997) Br J Med 315:81
66. López-Carrillo L, Blair A, López-Cervantes M, Cebrián M, Rueda C, Reyes R, Mohar A, Bravo J (1997) Cancer Res 57:3728
67. Hunter DJ, Hankinson SE, Laden F, Colditz G, Munson JE, Willett WC, Speizer FE, Wolff MS (1997) New Engl J Med 337:1253
68. National Research Council (1999) Hormonally-Active Agents in the Environment. National Academy Press, Washington, DC, p 6

57. [illegible] PV, Cohn [illegible] (1998) Fertil Steril 70:[illegible]
58. Becker S, [illegible] (1997) Fertil Steril [illegible]
59. [illegible] (1998) Nature [illegible]
60. [illegible] (1998) Chemosphere [illegible]
61. [illegible]
62. [illegible] (1998) [illegible]
63. [illegible] (1992) [illegible]
64. [illegible] (1994) [illegible]
65. [illegible]
66. [illegible]
67. [illegible]
68. [illegible]

CHAPTER 9

The Endocrine Disrupting Potential of Phthalates

Catherine A. Harris, John P. Sumpter

Department of Biological Sciences, Brunel University, Uxbridge, Middlesex, UB8 3PH, U.K.
e-mail: catherine.harris@brunel.ac.uk

Phthalate esters are ubiquitous in today's environment. Both terrestrial and aquatic organisms are subject to a low level but constant exposure to this class of chemicals. Until very recently, it was not thought likely that any phthalates would display endocrine activity, and hence very little, if any, research focused on this possibility. When reproductive effects were observed, they were not interpreted as being due to any intrinsic endocrine activity of phthalates (or their products of metabolism), but rather due simply to a "toxicity" of unknown mechanism. However, recently a small number of phthalates has been found to elicit estrogenic responses in *in vitro* assays. None of these, however, have been found capable of inducing specifically estrogen-dependent effects *in vivo*. It is unlikely that phthalates alone are responsible for what may be endocrine-mediated adverse effects observed in wildlife and humans over the past few decades, but it is possible that they are a contributory factor to this phenomenon. Phthalates administered in high doses to adult mammals have caused adverse reproductive development in their offspring. Recent thinking has proposed that these manifestations may be as a result of an anti-androgenic mechanism. This theory should be investigated in greater depth, and at environmentally relevant concentrations of the active phthalates. Before it is possible to assess the risks (if any) of exposure to phthalates, a much wider range of test species, and a wider range of endpoints, particularly endocrine ones, need to be assessed.

Keywords. Phthalate, Estrogenic, Anti-androgenic, Reproductive toxicity

1 **Chemistry and Uses** 171

2 **The Ubiquity of Phthalates** 173

2.1 Fate in the Environment 173
2.1.1 Phthalates in the Aquatic Environment 173
2.1.2 Phthalates in the Terrestrial Environment 174
2.1.3 Degradation of Phthalates in the Environment 175
2.2 Exposure 176
2.2.1 Exposure of Humans to Phthalates 176
2.2.2 Exposure of Aquatic Organisms to Phthalates 178
2.3 Metabolism and Bioaccumulation 178
2.3.1 Mammals 178
2.3.2 Aquatic Organisms 180

The Handbook of Environmental Chemistry Vol. 3, Part L
Endocrine Disruptors, Part I
(ed. by M. Metzler)

3 **Reproductive Effects of Phthalates** 181

3.1 Reproductive Toxicity 181
3.1.1 Mammals 182
3.1.2 Aquatic Organisms 184
3.2 Estrogenic Activity of Phthalates 186
3.2.1 *In Vitro* 187
3.2.2 *In Vivo* 190
3.3 Anti-Androgenic Activity of Phthalates 193

4 **Discussion** 195

5 **References** 198

Abbreviations

AR	androgen receptor
BBP	butyl benzyl phthalate
BCF	bioconcentration factor
DBP	dibutyl phthalate
DEHP	di-2-ethylhexyl phthalate
DEP	diethyl phthalate
DHP	dihexyl phthalate
DHT	dihydrotestosterone
DIBP	diisobutyl phthalate
DIDP	diisodecyl phthalate
DIHP	diisohexyl phthalate
DINP	diisononyl phthalate
DMP	dimethyl phthalate
DnHP	di-*n*-hexyl phthalate
DnOP	di-*n*-octyl phthalate
DnPP	di-*n*-propyl phthalate
DOP	dioctyl phthalate
E2	17β-estradiol
ER	estrogen receptor
FSH	follicle-stimulating hormone
LH	luteinising hormone
MAFF	Ministry of Agriculture, Fisheries, and Food (UK)
MBP	monobutyl phthalate
MBzP	monobenzyl phthalate
MEHP	mono-2-ethylhexyl phthalate
MEOH	methanol
MnHP	mono-*n*-hexyl phthalate
MnOP	mono-*n*-octyl phthalate
MnPP	mono-*n*-propyl phthalate
mRNA	messenger ribonucleic acid
4-NP	4-nonylphenol

OECD	Organisation for Economic Cooperation and Development
PVC	polyvinyl chloride
TDI	tolerable daily intake
Zrp	zona radiata proteins

1 Chemistry and Uses

Phthalate esters are among the most widely used industrial chemicals in existence. They are used principally as plasticisers, to impart flexibility, workability, and durability to polymers, but they can also be found in products such as paints, adhesives, inks, and cosmetics. Millions of tonnes of phthalates are produced world-wide every year, with hundreds of thousands of tonnes being used in Europe alone (see Table 1). The significance of phthalate consumption lies in the fact that their applications are widespread and extremely diverse. This, together with the fact that their very nature determines that they are fluid within the materials to which they are added and thus can leach from these media, leads to the ubiquitous presence of phthalates in environmental samples. On the positive side, however, phthalates are far more reactive than many other industrial organic contaminants with low solubilities, and therefore degrade more easily in the environment; thus although the input is constant, the final concentration may be lower than initially expected. Consequently, major questions that arise are "to what do they degrade?", and "are the intermediate degradation products harmful in themselves?"

Table 1. Use of phthalates. Figures are given for the European consumption of the most widely used phthalates. Note that this information will differ from the consumption of phthalates in other continents. Values of water solubility and log K_{ow} are taken as recommended by Staples et al. [6], indicating the extremely low solubility of this class of chemicals, particularly those with longer side-chains

Name	Abbr.	Molecular weight	Water [6] solubility (mg/L)	log K_{ow} [6]	Mass consumed in Europe (tonnes annum^{-1})
Di-2-ethylhexyl phthalate	DEHP	390	0.003	7.50	400,000–500,000
Diisononyl phthalate	DINP	425	<0.001	>8.0	100,000–200,000
Diisodecyl phthalate	DIDP	447	<0.001	>8.0	100,000–200,000
Butyl benzyl phthalate	BBP	312	2.7	4.59	20,000–50,000
Dibutyl phthalate	DBP	278	11.2	4.45	20,000–40,000
Diisobutyl phthalate	DIBP	278	20.0	4.11	20,000–40,000
Ditridecyl phthalate	DTDP	525	<0.001	>8.0	3,000–10,000
Diethyl phthalate	DEP	222	1100	2.38	(with DMP) 10,000–20,000
Dimethyl phthalate	DMP	194	4200	1.61	(with DEP) 10,000–20,000

DBP &
DIBP(*)

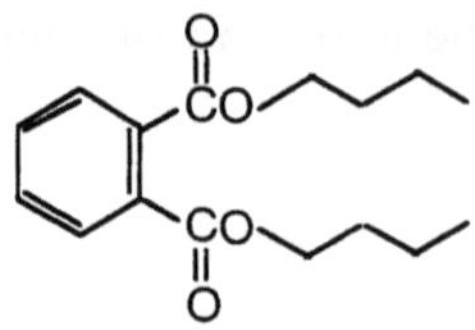

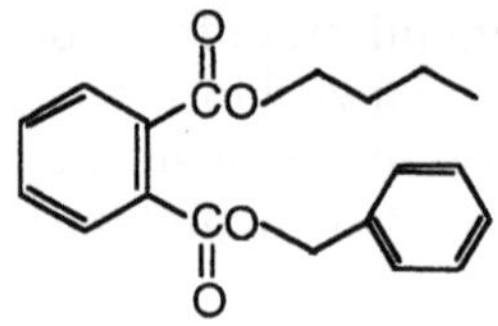

BBP

DEP

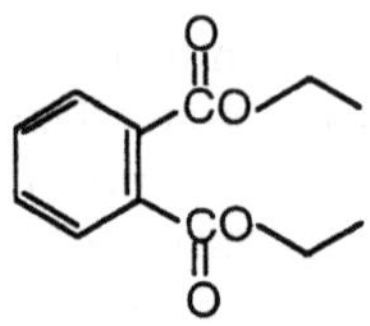

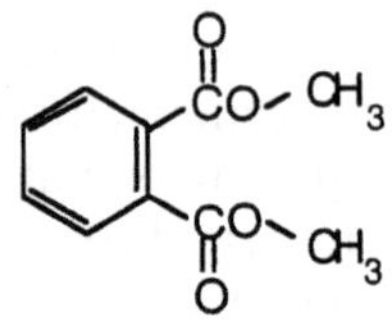

DMP

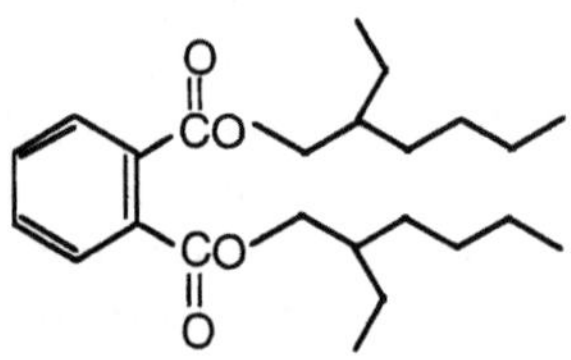

DEHP

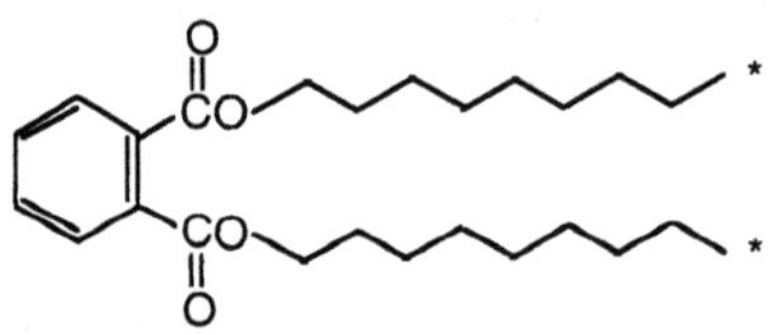

DINP

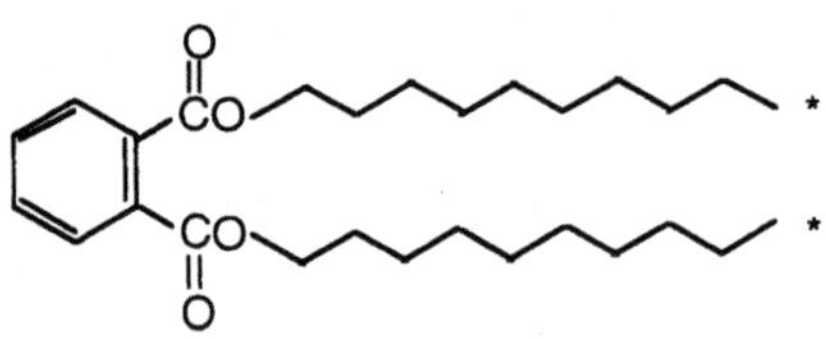

DIDP

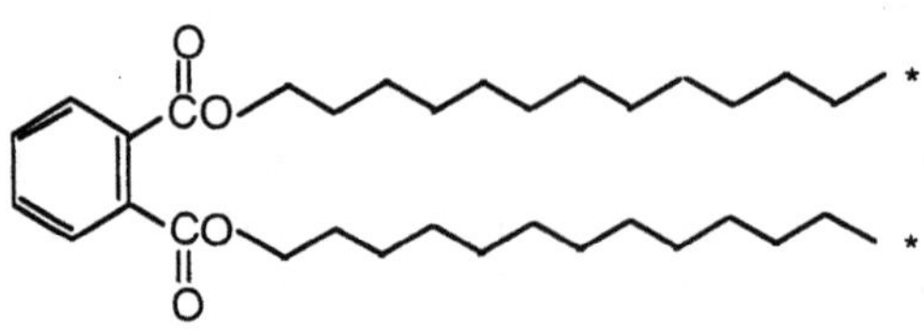

DTDP

Fig. 1. Structures of the phthalates which are most widely consumed in Europe. An asterisk indicates that the side chains may be branched, such that several isomeric forms of the phthalate exist

The basic structure of a phthalate ester comprises an aromatic ring with 2 aliphatic side chains. There are more than 60 different phthalates in use. Generally, however, only a small number is used in any large quantity in Europe; these are listed in Table 1, and their structures are shown in Fig. 1. As a class, the popularity of phthalates as plasticisers is owed to their inertness, fluidity, and high solubility in the polymer [1]. Also significant are their low water solubilities, and low volatilities, which help prevent loss from the polymer and thus ensure the continued flexibility of the plastic. The lengths of the side chains determine the properties of each individual phthalate, and therefore its end use.

Di-2-ethylhexyl phthalate (DEHP) is by far the most widely employed phthalate, and hence much of the research into the behaviour and toxicity of this class of chemicals has focused on DEHP and its metabolites. The major end use of DEHP is as a plasticiser in polyvinyl chloride (PVC), in which it can frequently be found forming up to 40% of the end product. Applications of this type of PVC include such human contact materials as childrens' toys and blood bags. Other phthalates used in large volumes that will also feature prominently in the following sections of this chapter include butyl benzyl phthalate (BBP), which is primarily used to make vinyl floor tiles, but has alternative applications, for example, in car upholstery and adhesives, and dibutyl phthalate (DBP), used primarily as a plasticiser, but which also has a wide variety of non-plasticiser applications, such as an ingredient in paints, inks, glue, nail polish, hair spray, and insect repellents [2].

Diisononyl phthalate (DINP) is also currently under investigation, not only because it is a high volume use phthalate, but also because a number of recent studies have discovered significant concentrations (constituting up to 50% of the product) in childrens' toys, many of which are designed specifically for sucking by young children; for example, teethers [3,4]. Since very young children are known to be at a very sensitive stage of development, it is important to know that they are safe from any chemicals to which they may be exposed, be it in low doses chronically, or in a high acute dose. There are currently many calls from consumer groups to ban the use of phthalates in toys, and several retailers already have a voluntary commitment to removing such products from their shelves. Hence, the use pattern of phthalates as discussed here may well alter in the near future.

2
The Ubiquity of Phthalates

2.1
Fate in the Environment

2.1.1
Phthalates in the Aquatic Environment

Phthalates can enter the aquatic environment via a number of pathways. These include in industrial waste direct from processing plants, in leachate from disposal of end products, in domestic and industrial sewage effluents, and in waste from vehicle washing.

It is generally accepted that biodegradation of phthalates in water is relatively fast (the half life obtained from shaking flask studies being reported as a matter of days, as opposed to months or years for some chemicals, [5]), and therefore measured concentrations of the parent compounds are, on the whole, in the range of the low μg L^{-1} to undetectable. However, this does not rule out accumulation of phthalates in the often anaerobic sediments of aquatic environments, which may lead to exposure of sediment dwelling or feeding organisms. Indeed, the K_{ow} values calculated for phthalates (ranging from 7.5 for the longer chain DEHP, to 4.45 for DBP, and 1.61 for the short chain dimethyl phthalate (DMP); [6]), suggest that phthalates may have a tendency to bind to particulate matter, which will settle out to form sediment. This is supported by the data of Thuren [7], who detected DEHP concentrations of 1.2 to 628 mg kg^{-1} dry weight sediment (d.w.) in the river Ronnebyan (Sweden), whereas concentrations in the water were 0.32 to 3.1 μg L^{-1}. Likewise, van der Velde et al. [8] measured concentrations of DEHP and DBP associated with particulate matter of 0.6 to 27 mg kg^{-1} d.w., and 0.1 to 1.5 mg kg^{-1} d.w., respectively, whereas the corresponding concentrations in the water phase were much lower, being at 0.6 to 7.5 μg L^{-1} and 0.2 to 0.7 μg L^{-1}. More extreme differentials were recently discussed by Long et al. [9], who detected concentrations of DEHP in water around the low μg L^{-1} mark, with the highest measurement being 21 μg L^{-1}. In the same rivers, concentrations of up to 115 mg kg^{-1} were measured in suspended sediment samples, with a general trend for higher concentrations of phthalates in suspended sediments than in bed sediments being observed. The concentration of DEHP in the sediments of all but two samples was in the low mg kg^{-1} to the tens of mg kg^{-1} range, therefore being at least 3 orders of magnitude higher than was measured in the corresponding water samples. Nonetheless, phthalates have been found in all compartments of the aquatic environment; for instance, Giam et al. [10] found them in water, sediment, air, and biota sampled from the Gulf of Mexico, and they have also been reported in marine samples taken from the North and Irish seas [11]. Freshwater studies have also revealed detectable concentrations of phthalates in Italian [12] and Malaysian [13] water and sediment samples, and in river water samples taken in the U.K. [14], Nigeria [15], and the U.S.A. [16]. Higher concentrations are frequently reported in samples from less developed countries, a fact that has been attributed to lack of legislation and/or facilities to treat industrial effluents [13, 15].

Sewage effluent samples have also been found to be contaminated with phthalate residues. Concentrations of DEHP up to 245 μg L^{-1} were found in Scottish effluents [17]. In that particular study, DEHP was detected in 59% of samples taken. Other phthalates, such as DEP, DBP, DIBP, and BBP were detected, but at reduced frequency and far lower concentrations.

2.1.2
Phthalates in the Terrestrial Environment

Phthalates have also been reported in samples taken from various compartments of the terrestrial environment, including the atmosphere, sludges, soils, landfill sites, and plants.

One route of entry of phthalates into soil systems is via fallout from the atmosphere, as discussed by Thuren and Larsson [18]. Their research led to an estimation of fallout of 202 μg m^{-2} DBP and 285 μg m^{-2} DEHP per annum to Swedish soils. They also found a temperature-dependent profile of phthalates in atmospheric samples, whereby lower concentrations were detected in air sampled in winter than in summer. Mean concentrations of DEHP and DBP were in the low ng m^{-3} range, similar to those detected by Giam et al. [19] in atmospheres of the Gulf of Mexico. The latter study found DEHP and DBP to be associated with both the particulate and the vapour phase in the air.

Another potential route of entry is via the application of sewage sludge to soil. Large volumes of sludge are disposed on to agricultural land, providing beneficial soil fertilising and conditioning properties. It is therefore of significant interest to determine whether potential toxins in the sludge might find their way into crop plants grown on this land. Kirchmann and Tengsved [20] investigated the uptake of phthalates by barley grown on soil amended with sludge and with pig slurry. The data obtained in these experiments revealed significantly higher concentrations of DEHP and DBP in grains obtained from barley grown on land amended with sludge and pig-slurry, respectively, than in control crops.

A summary of the properties of individual phthalates controlling their fate in the environment (such as solubility, K_{ow}, and vapour pressure) can be found in Staples et al. [6].

2.1.3
Degradation of Phthalates in the Environment

Despite the constant input and widespread distribution of phthalates into environmental systems, the rapid degradation of these chemicals is of paramount importance in preventing, in most instances, their accumulation. As a rule, biodegradation pathways are dominant in surface waters, soils, and sediments, whereas photodegradation prevails in the atmospheric compartment [6].

The degradation rates of phthalates are dependent on their molecular weight; those with longer alkyl side-chains tend to have longer half-lives in a given environment. This phenomenon was observed by Shelton et al. [21], who found DnOP and DEHP to remain intact in anaerobic digester sludge, whereas the lower molecular weight DMP, DEP, and DBP were completely mineralised, as was BBP, albeit at a slightly reduced rate. A similar picture was reported by Ejlertsson et al. [22, 23], who found that more soluble phthalates were degraded in anaerobic conditions, whilst less soluble esters were not.

In aerobic environments, e.g., activated sludge systems [24], or the modified Sturm test [25], the higher molecular weight phthalates can also be transformed, albeit at a slower rate than those with shorter alkyl side-chains.

Biodegradation in soils of DBP [26] and DEHP [27] has also been reported. Jianlong et al. [26] found that 76% of DBP was broken down in a natural (unsterilised, non-inoculated) soil sample after 30 days. Roslev et al. [27] investigated degradation of DEHP in a sludge-amended soil and found the rates to be dependent on the bioavailability of the chemical. Thus, although initially the half-life of DEHP was 58 days, it was calculated that 40% would remain after a one year incubation, due to lack of bioavailability.

With respect to aqueous degradation, Group [28] observed a correlation between biodegradation half-life and length of alkyl side chains in shake flask and activated sludge studies, but concluded that all are 'readily degraded'. Saeger and Tucker [29] also reported rapid degradation of phthalates such as BBP and DEHP in river water and activated sludge, but they began their studies using concentrations much higher than those commonly found in surface waters (1 to 83 mg L^{-1}). Microcosm studies by Adams et al. [30] revealed rapid primary degradation of BBP (with a half-life of less than 2 days) and estimated the rate of mineralisation to be relatively fast also (with a half-life of 13 days). Biodegradation studies with phthalates in river water were carried out by Furtmann [31] who found that, as a rule, more than 90% was degraded within 5 days, but once more, DEHP was slower to break down, particularly at concentrations of less than 2 µg L^{-1}. As might be expected, degradation in these laboratory conditions was slower at 4 °C than at 20 °C, implying that phthalate degradation in the environment will be slower in winter than in summer.

It can be concluded that there is a general trend for decreasing biodegradability of phthalates corresponding to increasing molecular weight. Hence, DEHP, the most prolific phthalate in use today, is also one of the most persistent in the environment [32].

2.2 Exposure

2.2.1 *Exposure of Humans to Phthalates*

The majority of studies discussing exposure of humans to phthalates involve oral exposure, for instance, through ingestion of food, or the chewing of toys by small children. Other potential routes may include skin contact, for example, from constantly handling plastic articles, contact with PVC clothing or vinyl childrens' pants, and cosmetics; or inhalation, for example, as a result of sitting in a car where phthalates are volatilising from upholstery into the atmosphere of the car interior. However, due to the difficulties associated with quantifying such exposure, there are no actual data on the degree of exposure via these pathways; therefore, whilst they must be borne in mind, the attention here will focus upon oral exposure of humans to phthalates. It must also be considered that any exposure estimates are purely estimates, and too much weight should not be placed on such subjective analyses.

Before we consider oral exposure, one other route must be mentioned: that arising from the use of phthalates for medical equipment, which may lead to intravenous exposure following, for instance, the storage of blood in PVC blood bags. The phenomenon of migration of DEHP into blood stored in this way was reported by Jaeger and Rubin [33], but the migration of DEHP from medical devices had been discussed as early as 1960 (cited in [33]). Other medical equipment found to leach phthalates directly into humans includes dentures. Lygre et al. [34] found DBP in saliva samples of patients who had recently received new dentures.

As previously mentioned, recent studies have highlighted the significant proportion of phthalates, mainly DINP, but to a lesser extent DEHP, DIDP, DEP, DBP, and BBP, in children's toys. DINP and DEHP can in many cases make up 10 to 40% by weight of PVC toys, including teethers which are intended to be put into the mouths of small children [3, 4]. These studies established concentrations of phthalates in the toys, but not their leaching potential. This latter question has been addressed by Steiner et al. [35] and Vinklesoe et al. [36]; the former found DEHP leaching from PVC child articles into a saliva simulant, and the latter reported the leaching of a range of phthalates from PVC teethers. These data demonstrate a real possibility of exposure of young children to phthalates from PVC toys. Bearing this in mind, there has been a long-running dispute between consumer groups and PVC manufacturers, leading in 1996 to several U.K. retailers grouping together to investigate the environmental and health effects of PVC [37]. This was followed by the removal of PVC baby products by Scandinavian retailers in May 1997 [38], and more recently by efforts within the EC to impose restrictions on the use of PVC toys containing phthalates [39]. On the other hand, the US Consumer Product Safety Commission (CPSC) in December 1998 advised that there is little risk to children from DINP, since the amount they are likely to ingest does not reach a level that would be harmful [40]. In the same report, however, they requested the removal of phthalates from soft rattles and teethers whilst awaiting further studies. There appears to be a great deal of confusion within both the retailer and consumer worlds as to the right course of action. This is not surprising, since measurements of exposure are themselves imprecise, and because the actual toxicity of phthalates is uncertain; that is, it is unknown whether the level of exposure obtained from childrens' toys could be responsible for any adverse effects.

Exposure of humans to phthalates via food is also a potential risk, as is evidenced by the number of studies which have analysed phthalates in food packaging materials [41–43], the migration of phthalates from such packaging [44], and the subsequent concentrations of various phthalates in food products [45–50]. It should be made clear, however, that some researchers consider increased phthalate concentrations to arise not only from packing materials, but also from general environmental contamination during, for example, the processing of the food products [45, 48]. Further, the British Plastics Federation (BPF) has stated that phthalates are no longer used in PVC food packaging or wrapping which is manufactured in the U.K. [37].

Nonetheless, phthalates have been detected in food samples, and this has been attributed to migration from materials such as plasticised lid seals, pie carton windows, and aluminium foil-paper laminated wrapping. Since dairy products are often fatty in nature, the potential for phthalate migration into these goods is fairly high. Concentrations of total phthalates as high as 114 mg kg^{-1} were found in cheese samples in the U.K. [48], and Page and Lacroix [50] reported concentrations of up to 11.9 mg kg^{-1} in butter. Human exposure to phthalates from food was estimated at 0.1 to 0.8 mg $person^{-1}$ day^{-1} (total phthalate) for an average person, and 0.4 to 1.6 mg person^{-1} day^{-1} for a high level intake [45]. The lowest tolerable daily intake (TDI) as recommended by the EC Scientific Committee for Food [51] for an individual phthalate (DBP or DEHP, or DIDP) is set at 3 mg person^{-1} day^{-1}; TDIs for other phthalates have been set at higher levels [51]. Therefore, considering that the concentrations estimated by MAFF [45] are more likely to be overestimates than underestimates, and that these figures are calculated for total phthalate consumption, the estimated exposure is well within the TDI levels set.

Young children are at a particularly sensitive developmental stage, and surveys have been carried out on infant milk formulae to assess phthalate contamination. An initial survey, in 1996 [46], found concentrations of 1.2 to 10.2 mg kg^{-1}, prompting newspaper headlines such as "Sex-change chemicals in baby milk", and "Doctors under siege in baby milk scare", but a follow-up survey [47] has reported reassuringly lower concentrations in similar samples, of <0.1 to 0.6 mg kg^{-1}, thereby lowering the overall exposure of infants fed on these diets to within set safety limits.

The exposure of humans to phthalates is unquantifiable owing to the ubiquity of these chemicals, and the considerable number of potential pathways, many of which scientists have not even attempted to assess. Therefore, it is not possible to state that the apparently low exposures from food products, however encouraging they are, indicate an intake which is safe overall.

2.2.2
Exposure of Aquatic Organisms to Phthalates

There is a distinct lack of research specifically devoted to exposure of aquatic organisms to phthalates. Briefly, however, pathways may include: intake via the gills of fish, uptake through food sources (which may themselves have absorbed phthalates from the water, or ingested food which had done so), intake of suspended sediments, and uptake due to living or feeding in the sediments of the water body floor. Uptake via the gills of fish is potentially an especially direct source of contaminant to the bloodstream, because the blood reaches many organs before passing through the liver, where most metabolism will occur, and thus this route of exposure should be researched more fully.

2.3
Metabolism and Bioaccumulation

2.3.1
Mammals

Metabolism of environmental contaminants can have major implications when considering their mechanisms of toxicology. *In vitro* assays seldom, possibly never, display the same metabolic pathways as whole organisms, and hence, if any effects of toxicants (such as phthalates) are due to metabolites, rather than the parent compounds, then only *in vivo* studies can provide the necessary information; results of *in vitro* assays, by themselves, may be misleading.

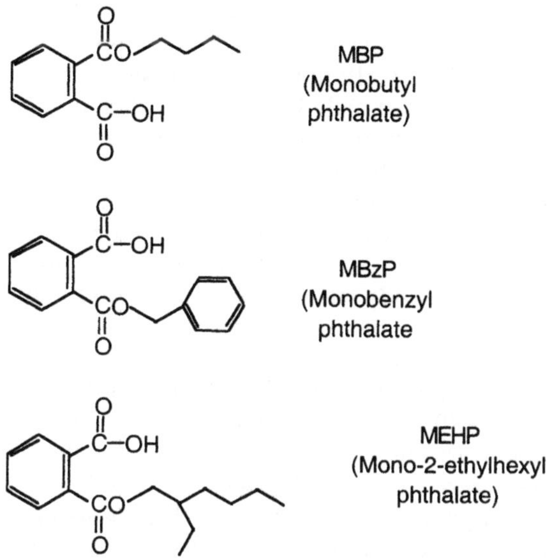

Fig. 2. Examples of the structures of the primary metabolites of some of the phthalates used in high volumes: MBP (primary metabolite of DBP and BBP); MBzP (also a primary metabolite of BBP – either MBP plus benzyl alcohol, or MBzP plus butyl alcohol, would be produced); MEHP (primary metabolite of DEHP)

The primary metabolites of phthalates are their monoesters. The first step in phthalate degradation is hydrolysis, whereby one of the side chains is cleaved, leaving the monoester plus an alcohol group. Examples are shown in Fig. 2.

The literature covering degradation of phthalates in mammals is, on the whole, limited to DEHP, of which there is a wealth of information. In addition, the majority of studies have focused on oral or intravenous (i.v.) routes of exposure. Exceptions include recent studies by Dirven et al. [52, 53], which discussed the metabolites of DEHP found in the urine of workers in the PVC industry. In these studies, figures are given showing concentrations of DEHP in the air of the working environment, so it can be assumed that the major route of exposure here is inhalation. The researchers analysed the four main metabolites of DEHP: namely mono-(2-ethylhexyl) phthalate (MEHP); mono-(5-carboxy-2-ethylpentyl) phthalate (metabolite V); mono-(2-ethyl-5-oxohexyl) phthalate (metabolite VI); and mono-(2-ethyl-5-hydroxyhexyl) phthalate (metabolite IX). Workers in a PVC boot factory were found to excrete increased concentrations of all four metabolites over the period of a work shift [52], thus implying that the occupational exposure to DEHP of workers in these industries can be assessed by monitoring metabolite concentrations in their urine. Dirven et al. [53] also investigated the chemical status of the metabolites, and discovered that metabolites VI and IX were mainly excreted in conjugated form, whereas metabolite V was found mainly in the free form, and that the degree of conjugation of MEHP varied between individuals.

Ng et al. [54] studied absorption and metabolism of DEHP through the skin of the hairless guinea pig. This is also considered to be a major route of occupational exposure. They concluded that most of the DEHP administered to the animals actually remained in the skin, due to the lipophilicity of the chemical, and that the metabolism of DEHP in this tissue is slow.

Oishi and Hiraga [55] dosed male Wistar rats with a high dose (9.75 g DEHP kg^{-1}) and monitored concentrations of MEHP and DEHP in body organs and adipose tissue. The longest half-life of DEHP was found to be in adipose tissue (at 156 hours), where the concentration of this chemical was higher than that of its monoester. In all other organs studied, and in blood, the concentration of MEHP exceeded that of DEHP. This might lead to conclusions that DEHP has the potential to accumulate in body fat. However, Schulz and Rubin [56] concluded that at high doses of DEHP (200 mg kg^{-1} administered intravenously), hepatic uptake processes which remove DEHP from the bloodstream reached a saturation point, therefore increasing the half-life of DEHP at these concentrations. In contrast, at lower doses (0.1 mg kg^{-1}), the initial rate of disappearance of DEHP from the bloodstream was much higher, and almost all of the DEHP from this intravenous dose was excreted as metabolites within 24 hours. Put into context, even this lower dose is greater than the estimated exposure of an average human to total phthalates via diet, as calculated by the U.K. Ministry of Agriculture, Fisheries and Food (0.013 mg kg^{-1} body weight day^{-1}) [45]. When the same researchers administered 200 mg DEHP kg^{-1} orally to rats, the rate of metabolism and excretion appeared to be faster than the rate of absorption, such that DEHP was not found in significant quantities in blood or tissues sampled.

Some analysis has been carried out on metabolites of DBP also. For example, Saillenfait et al. [57], whilst conducting a study on the developmental toxicity of DBP, also looked at degradation products. They found that DBP, administered to pregnant Sprague-Dawley dams, was metabolised to its monoester, monobutyl phthalate (MBP), and MBP glucuronide, the former of which accounted for the majority of the radioactivity recovered from maternal plasma, placenta, and embryonic tissue. DBP was not found to accumulate in any form in the dam or in the embryo. Radiolabelled DBP has also been fed to cattle, followed by analysis of bile, faeces, plasma, and urine [58], and MBP found to be a universal metabolite of DBP. Other more minor metabolites found in this study were phthalic acid, monoethyl phthalate, and monohydroxybutyl phthalate.

It must be recognised that extrapolations between varying experimental conditions should not be made automatically, as there are some considerable variations between metabolic capabilities of different species, and differences in metabolic rate and half-lives have been observed when using high versus low doses of DEHP.

2.3.2
Aquatic Organisms

A number of different species have been employed to assess the metabolism of phthalates by aquatic species. Wofford et al. [59] carried out a comprehensive survey of the metabolism of a number of phthalates in different organisms, leading to the conclusion that rates of biodegradation were more dependent on species than on either phthalate chain length or concentration. As observed in mammals, they found the primary metabolites of phthalates to be the monoesters, with a number of unidentified polar metabolites also being produced.

In algae (*Chlorella pyrenoidosa*), phthalates were observed to be accumulated, but the algae were also found to have metabolic capabilities [60]. In this species, biodegradation (and accumulation) depended on chain length; average degradation rates per day were 13.4 mg L^{-1}, 7.3 mg L^{-1}, and 2.1 mg L^{-1} for DMP, DEP, and DBP, respectively. Likewise, the bioconcentration factor (BCF), although time-dependent, reached maxima of 162 (after 24 hours), 205 (after 12 hours), and 4077 (after 12 hours) for DMP, DEP, and DBP, respectively [60].

Stalling et al. [61] exposed channel catfish to DEHP and found MEHP to be the major metabolite, with 66% of DEHP converted to MEHP after a 24-hour exposure to 1 µg DEHP L^{-1}. In fathead minnows, the ratio of MEHP:DEHP was lower, but the concentrations of DEHP and exposure period was different, so the experiments are not comparable. In the same study, the authors used hepatic microsomes to examine metabolism of DBP and DEHP. Degradation of DBP reached 97% after 2 hours, whereas only 6% of DEHP was metabolised in the same length of time.

Metabolism of DEHP in rainbow trout was examined by Barron et al. [62]. They found no metabolites that were unique to rainbow trout, but fewer oxidised metabolites than in mammalian species. As observed in mammalian studies, hydrolysis of DEHP to MEHP was the first step of degradation, and the phthalate ring itself was not oxidised.

In the bluegill sunfish, metabolism of BBP was studied [63]. The BCF for intact BBP was found to be 9.4 (for the whole fish). This was compared to estimated BCF values calculated by 4 different groups of scientists, using physical properties such as K_{ow} and solubility, of 3174, 2528, 705, and 304 (cited by [63]), which appear to be vast overestimates. The degradation products of BBP were not characterised by Carr et al. [63].

Despite these studies, our knowledge of metabolism of phthalates in aquatic organisms is relatively poor. It is dangerous to generalise from the limited, and fragmented, information available to all aquatic organisms, both plants and animals, the latter ranging from invertebrates to higher vertebrates, which inhabit the variable aquatic environment.

3 Reproductive Effects of Phthalates

3.1 Reproductive Toxicity

Literature covering the reproductive toxicity of the major use phthalates is extensive, though mainly confined to mammalian studies. Unfortunately, many experiments have been carried out using very high concentrations of phthalates, to which it is unlikely that an average human would be exposed. Nonetheless, it is worth discussing a selection of these reports, if only to highlight the variety of adverse effects which phthalates are capable of eliciting in exposed organisms.

3.1.1
Mammals

Testicular toxicity of phthalates has been well documented; the first report covering this topic was in 1945 (cited in [64]). Subsequently, several authors have published reports of effects of phthalates on testicular weight and histology; a summary of many of these can be found in Gangolli [64]. Essentially, they describe decreased testicular weight and seminiferous tubular atrophy induced by DEHP, DBP, di-*n*-propyl phthalate (DnPP), and di-*n*-hexyl phthalate (DnHP), as well as by their respective monoesters. One important observation is that, as with the metabolism of phthalates described above, the response appears to be species-specific. For example, the testicular toxicity observed in the rat cannot be repeated in hamsters, or in testicular cell preparations from this apparently resistant organism. The effects described by Gangolli [64] were further explored by Gray and Gangolli [65], who confirmed the action of MEHP, MnOP, MnHP, MnPP, and MBP in causing the detachment of germ cells from Sertoli cells, which was consistent with the toxicity of their parent phthalates *in vivo*. With respect to DEHP in particular, it was found that MEHP was the most toxic metabolite, whereas DEHP and its other primary metabolite, 2-ethylhexanol, were not toxic *in vitro*, and metabolites V, VI, and IX were less toxic than MEHP. Thus, it was concluded that MEHP is the active testicular toxin produced as the result of metabolism of DEHP *in vivo* [65].

Studies are currently being undertaken by Hardell and colleagues to investigate the mechanisms behind their finding that the exposure of men to PVC led to an increased likelihood that they would develop testicular cancer [66]. The reason for this is not known, but was attributed to an ingredient specific to PVC, since exposure to other types of plastic did not increase the risk of this disease.

On a cellular level, the effect of phthalates on rat Leydig cells was investigated by Jones et al. [67]. The authors studied the activities of DEHP, DnPP, DnOP, and DEP *in vivo*, and their corresponding monoesters *in vitro*. The data acquired showed an effect on Leydig cell structure and function both *in vivo*, with the parent phthalates DEHP and DnOP, and *in vitro* with the monoesters MEHP and MnOP. DEP affected Leydig cells *in vivo*, but DnPP and MnPP exerted no effect in either assay. Richburg et al. [68] exposed young Fischer rats to MEHP and found that it adversely affected Sertoli cells, which in turn resulted in disruption of germ cell apoptosis. Recently, Li et al. [69] found that low concentrations (up to 1 μM) of MEHP induced the detachment of gonocytes (the precursors of spermatogonia) from Sertoli cells. For this work, an *in vitro* coculture of Sertoli cells and gonocytes was employed. The cells had been isolated from 2-day-old pups, thus underlining the sensitivity of neonatal testicular development.

Semen quality can also be affected by phthalates, as evidenced by Fredricsson et al. [70], who reported decreased motility of human spermatozoa exposed *in vitro* to DEHP and DBP. Pennanen et al. [71] found a slight decrease in fertility of female Wistar rats, and decreased sperm quality in males, exposed to 2-ethylhexanoic acid (a metabolite of DEHP). Decreased sperm count of rats exposed to DEHP was observed alongside a decreased epididymal weight by

Siddiqui and Srivastava [72]. Taking a slightly different approach, Murature et al. [73] plotted sperm density versus concentration of DBP in semen, and observed a negative correlation, although this analysis was based on a low number of subjects.

Modification of steroid concentrations in female rats has profound implications for their reproductive capability. For example, adult females exposed to DEHP experienced a decrease in serum estradiol concentrations [74]. This in turn led to feedback effects on gonadotropins, whereby follicle-stimulating hormone (FSH) levels were increased, but luteinising hormone (LH) concentrations were suppressed, such that the LH surge essential for ovulation did not occur. Consequently, there was a lack of ovulation in the rats dosed with DEHP. Laskey and Berman [75] also observed that DEHP altered the steroid profile of rat ovarian cultures. Similarly to the observations of Davis et al. [74], they found a decrease in estradiol concentrations of cells which had been stimulated such that they were 'in estrus'. In treated cells in diestrus, the concentrations of both testosterone and estradiol were increased.

Ema and colleagues have undertaken numerous studies to determine the developmental toxicity of phthalates to rats exposed *in utero*. In the process, they found that maternal exposure to BBP induced decreased uterine and ovarian weights, decreased progesterone concentrations, and that post-implantation embryonic loss was increased at day 11 of pregnancy [76]. The same scientists also found similar results when rats were exposed to DBP, although the decrease in progesterone levels was found to be less pronounced than that which occurred in the BBP studies [77]. MBP, a metabolite common to both DBP and BBP, also increased post-implantation loss, and at a slightly lower dose, although only one dose had been employed for the DBP and BBP experiments, so the effective concentrations in these studies cannot be compared to those observed in the MBP experiment [78]. More recently, Ema et al. [79] have reported studies whereby DBP has been administered to rats during the latter half of pregnancy, and effects (if any) on the offspring were investigated. DBP had an overall adverse effect with respect to the reproductive development of male fetuses. These effects included an increased number of fetuses with undescended testes, which confirms data presented by Imajima et al. [80], who demonstrated that MBP, fed to pregnant rats on days 15–18 of gestation, induced a significantly higher testicular ascent than could be observed in controls. This latter study also revealed cryptorchidism in 84.6% of 30–40-day-old rats (compared to 0% in the control group), suggesting that effects seen in the fetus, exposed whilst still developing, are carried through to the growing, but no longer exposed, young rat.

Studies performed using BBP have illustrated its general reproductive toxicity. Agarwal et al. [81] demonstrated a whole kaleidoscope of adverse reproductive effects in adult male Fischer rats. These ranged from decreased weights of epididymis, testis, prostate, and seminal vesicles, accompanied by atrophy of the latter three, through to decreased plasma testosterone concentrations, and an increase in FSH and LH levels. The conclusion of the authors was that BBP exerted a direct effect on the testis, with secondary effects occurring in other reproductive organs. These data were reflected in studies by Piersma et al. [82], who were evaluating the OECD 421 reproductive toxicity screening protocol. Their studies confirmed BBP as a reproductive toxicant, with influences on parameters such as testis and epididymis weights, spermatogenesis, time to conception, pregnancy rate, post-implantation survival, and litter size and weight.

An extremely significant study, highlighting a multigenerational effect, was published by Wine et al. [83]. In this study, rats of the F_0 and F_1 generations were exposed to DBP, and various developmental and reproductive parameters were assessed. When a mating trial was performed using the F_1 animals, it was found that these pups had decreased mating, pregnancy, and fertility indices, compared to control groups. In addition, the F_1 generation had decreased sperm counts, and a higher incidence of degenerated seminiferous tubules and imperfect epididymides. The F_0 generation, in contrast, had not suffered any adverse effects on sperm parameters or estrous cyclicity, although the number of live pups per litter, and the weight of the pups, were decreased by DBP. The authors concluded, therefore, that adverse effects observed in DBP-exposed pups of the second generation were enhanced compared to those seen in the first generation. Mechanisms of action were not evaluated in this study, but it was suggested that the effects observed may have been induced by an ER-mediated interaction, since they were similar to those observed in rats and mice which had been exposed to diethylstilbesterol (DES) [83]. However, the spectrum of effects in developing males exposed to either estrogens or anti-androgens is similar, as discussed by Sharpe [84], making it difficult to deduce the mechanism of action (there may be more than one mechanism, of course).

In summary, DEHP, DBP, BBP, and their monoesters, can all exert adverse reproductive effects on mammalian species. The mechanisms by which these effects occur have occasionally been investigated in detail, but it is the ultimate effect on reproductive capability which is important, as this could have an impact at the populational level of the organism concerned. In the majority of cases, as is normal for toxicity studies in general, the doses administered were well above those that would be expected in a natural environmental situation. However, it is also true that exposure of organisms to environmental pollutants such as phthalates, although at lower doses, is potentially constant, and covers multiple generations, whereas toxicity tests usually (although not always) involve relatively short-term exposure. The relatively rapid metabolism of phthalates is likely to prevent a significant build up of these chemicals in an organism, but nevertheless, ideally, we would like to see long-term, multigenerational studies replacing short-term, high-dose experiments. Unfortunately, these are extremely costly and time-intensive, and it is unrealistic to expect all toxicity studies to replicate natural exposure conditions. At the same time, it is difficult to know exactly how to interpret data which are derived from unrealistic exposure conditions.

3.1.2
Aquatic Organisms

In general, as with mammalian toxicity studies, concentrations of phthalates used in these experiments are not representative of real environmental situations. In fact, several aquatic toxicity experiments have been performed using concentrations of phthalates which are far higher than the water solubility of the respective phthalate. This in itself can create problems, owing to a certain amount of undissolved chemical in, or floating on the surface of, the water.

A popular freshwater aquatic species for toxicity testing is *Daphnia magna* (the common water flea), due to its ease of breeding and general husbandry, together with its short life cycle, allowing for reproductive effects to be monitored over a relatively short period of time. For similar reasons, fathead minnows (*Pimphales promelas*) are frequently used as a model freshwater fish species. Fathead minnows have the additional benefit of laying eggs onto a solid surface rather than into the general water body, so fecundity can be more easily monitored. Rainbow trout (*Oncorrhynchus mykiss*) have also been employed in several toxicity screening tests, owing to their commercial significance, as well as their extreme sensitivity to environmental influences.

DeFoe et al. [85] found DBP to have an adverse effect on reproduction of *Daphnia* at 0.64 mg L^{-1}. DBP and DnOP were also reported to have adverse effects on the fecundity of *Daphnia* at 1.8 mg L^{-1} and 1.0 mg L^{-1}, respectively, by McCarthy and Whitmore [86]. This concentration of DnOP is one example where the test chemical was applied well above its solubility limit (see Table 1). More recently, authors have attributed the toxicity of high concentrations of phthalates to *Daphnia* to surface entrapment (that is, that the toxic effects observed were due to physical, as opposed to physiological, processes) in the excess undissolved chemical [87].

Abernathy et al. [88] concluded that the limited solubility of the higher molecular weight phthalates was not sufficient to allow a critical body burden to be achieved when organisms are exposed at concentrations less than the solubility limit of the phthalate concerned. However, it does not necessarily follow that because the solubility of a chemical is low (for example, DEHP), that exposure of organisms in the aquatic environment to that chemical will be negligible. There is also the possibility that organisms will be exposed to phthalates via sediment or suspended matter, where concentrations of lipophilic chemicals are often higher, sometimes by orders of magnitude, than are found in the aqueous phase [9].

In a study performed by McCarthy and Whitmore [86], DBP affected the hatch rate of fathead minnows; 1.8 mg L^{-1} prevented eggs from hatching altogether, whereas 1.0 mg L^{-1} suppressed hatch rate and larval survival. Two mg DBP L^{-1} also decreased egg production in *Rivulus marmoratus*, a self-fertilising fish [89]. Viability of the eggs produced was diminished during the exposure of the parent fish to 1 mg L^{-1} and 2 mg DBP L^{-1}, but this effect was not maintained after exposure had ceased.

When DEHP was administered to water such that the final concentration was 0.502 mg L^{-1}, there was no effect on the hatchability, survival, or growth of rainbow trout over a 90-day embryo-larval test [85]. An extensive study covering exposure of Daphnids to 14 different phthalates, and rainbow trout to 6 phthalates, concluded that for the lower molecular weight esters (DMP, DEP, DBP, BBP), toxicity increases as solubility decreases, and survival of exposed species was as sensitive as reproduction. The higher molecular weight phthalates appeared to be less toxic; with these, survival was more sensitive than reproduction [87].

To summarise, very few studies have explored in depth the reproductive effects of phthalates on freshwater organisms, and no chronic toxicity studies have been reported for saltwater fish to our knowledge. Aquatic toxicity of phthalates has been summarised by Staples et al. [90]. In brief, the higher molecular weight phthalates have, on the whole, not been found to be toxic to aquatic organisms, other than at concentrations above their solubility; the toxicity observed in these cases is thought to be due to the physical effects of undissolved phthalate. The toxicity data on the lower molecular weight phthalates (of which that with the highest molecular weight is BBP), are generally in agreement with that found by Rhodes et al. [87], whereby toxicity increases with increasing molecular weight. However, given the low water solubility of the phthalates, and therefore the fact that a large proportion of phthalates in the aquatic environment are likely to partition into the particulate phase, it is perhaps surprising how little we know about the chronic toxicity of these chemicals to sediment-dwelling organisms. We cannot therefore draw any robust conclusions on the overall effect of phthalates in aquatic systems.

3.2 Estrogenic Activity of Phthalates

Essentially, the fact that some phthalates are weakly estrogenic *in vitro* is now widely accepted. Potencies of phthalates in various *in vitro* assays are shown in Table 2. More complex is their activity in vivo, of which there are conflicting reports, as detailed below.

Table 2. Estrogenic potencies of phthalates (expressed as orders of magnitude *less* potent than E2) in different types of assays are shown as reported by the author (a); or have been calculated using IC_{50} values reported (b), using the formula $IC_{50}(x)/IC_{50}(E2)$. "Positive" indicates that the phthalate induced an estrogenic response, but potency relative to E2 could not be calculated

Phthalate name	Yeast-based assay	Mammalian cell line	Receptor binding
BBP	negative [94] 1×10^6 [97] (a)	positive [91] 3×10^5 [93] (a) positive [97]	positive [91] 2.8×10^4 [92] (b) 7×10^3 [99] (b) 2×10^5 [100] (a)
DBP	negative [93] positive [95] 1×10^7 [97] (a)	positive [91] positive [97]	positive [91] 3.6×10^4 [92] (b) negative [95] 2×10^5 [100] (a)
DIBP	1×10^7 [97] (a)	positive [97]	
DEHP	negative [95] negative [97]	negative [97]	positive [91] negative [95] 2×10^5 [100] (a)
DEP	5×10^7 [97] (a)		2×10^5 [100] (a)

3.2.1
In Vitro

The first report of estrogenic activity of phthalates was published as recently as 1995. Jobling et al. [91] assayed a number of xenobiotics in several *in vitro* assays, and found BBP, DBP, and DEHP to compete with 17β-estradiol (E2) for binding to the rainbow trout estrogen receptor (ER), albeit weakly so. That is, very high concentrations were required, and even at the highest concentration tested, the response was far less than the maximal response elicited by E2 itself. The same scientists investigated these phthalates for their ability to induce proliferation of ZR-75 breast cancer cells – a response specifically stimulated by estrogens. In that assay, only BBP and DBP were found to be estrogenically active, and again, their responses were less than maximal. This latter assay is a whole cell assay, and therefore has potentially increased metabolic capabilities (compared to a receptor binding assay), which may account for the loss of activity of DEHP in this assay. However, a subsequent report on the activity of phthalates in an estrogen receptor binding assay [92] also reported DEHP to be inactive. The third assay performed by Jobling et al. [91] assessed the abilities of the chemicals to stimulate transcriptional activity of the human estrogen receptor. In this assay, as in the former two, BBP was found to be the more potently estrogenic phthalate, and was active at concentrations of 1×10^{-6} to 1×10^{-4} M. DBP was active at 1×10^{-5} to 1×10^{-4} M, and DEHP was only active at concentrations exceeding 1×10^{-4} M. This pattern of estrogenic activity of these phthalates has been repeated in the majority of studies reporting the results of *in vitro* assays published since that of Jobling et al. [91]. There are some exceptions, such as data from Soto et al. [93], who concluded that of all the phthalates tested in the "E-Screen" (monitoring the proliferation of MCF-7 breast cancer cells), only BBP was estrogenic. These authors concluded that the alkyl phthalates were not estrogenic, although it was not clear at what concentrations they had been tested, and therefore it may be that the concentration of estrogenic alkyl phthalates used were simply not high enough to stimulate a response.

However, not all reports of the estrogenic activity of phthalates *in vitro* are in agreement. For example, Gaido et al. [94] found BBP, at concentrations up to $1 \times 10{-}^{4}$ M, to be inactive in a yeast-based assay. An assay to test for androgenic activity was also reported in that study, and BBP was found not to be an androgen agonist.

Petit et al. [95] also demonstrated the use of an *in vitro* yeast-based assay, in which the yeast had been transformed with a gene coding for the rainbow trout estrogen receptor. DEHP was not estrogenic in this assay, but DBP was weakly active. Both of these phthalates were inactive in a competitive binding assay based on the rainbow trout estrogen receptor, and DBP was found not to induce vitellogenin mRNA production (an estrogen-specific response) in hepatocyte cell cultures [95].

Another recombinant yeast-based assay, this time with the yeast transformed with a gene for the human ER, found BBP to be estrogenically active, but DBP and DEHP were inactive [96]. In this study, BBP was also tested in an *in vivo* assay, namely the mouse uterotrophic assay (in which the endpoint is relative

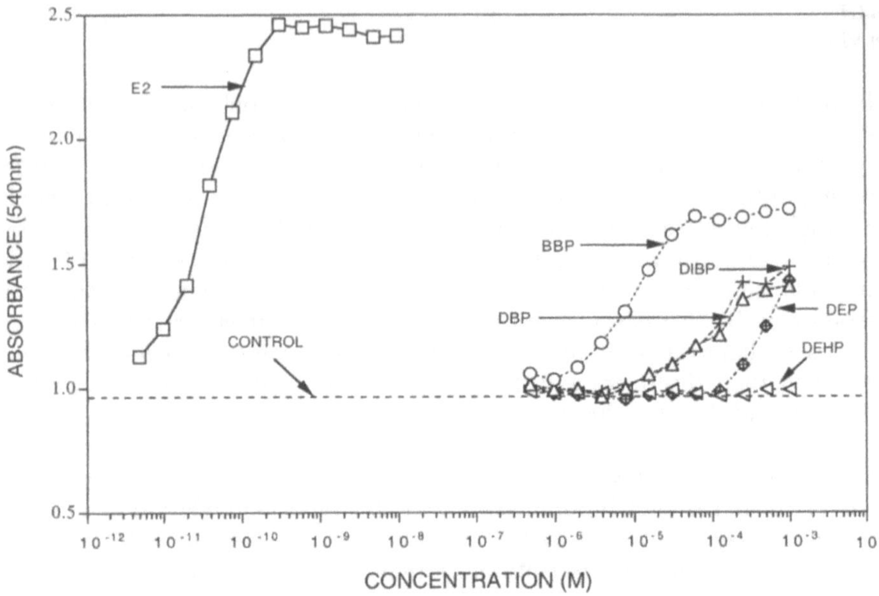

Fig. 3. The *in vitro* estrogenic activities in a recombinant yeast-based assay of some of the major volume use phthalates, each serially diluted from 10^{-3} M, as reported by Harris et al. [97]. The response to 17β-estradiol (E2), serially diluted from 10^{-8} M, is shown as a positive control

uterine weight, an increase in uterus weight indicating an estrogenic stimulus) and found to be inactive.

A comprehensive assessment of the estrogenic activity of phthalates (over 30 were tested) in different *in vitro* assays for estrogenic activity was reported by Harris et al. [97]. In this study, again only a partial response was observed for those phthalates which were found to be estrogenically active in a recombinant yeast-based assay (Fig. 3). The potency of these was of the order BBP>DBP> DIBP>DEP. DEHP was inactive in this study, and DINP showed extremely weak activity in some instances, but this was not repeatable. Some of the monoesters of the major phthalates (MBP, MBzP, MEHP, and MnOP) were also assessed for their estrogenic activity, along with metabolites V, VI, and IX of DEHP. All were inactive. These data provide a potentially important message; metabolites of phthalates have rarely been investigated for their endocrine activity *in vitro*, and yet it is perhaps as important, if not more important, to assess the potential of these chemicals to disrupt the endocrine system, due to the relatively rapid metabolism of the parent esters *in vivo*. Many *in vitro* assays, in particular receptor binding assays, which are widely employed in these estrogenicity screens, do not possess the full metabolic capabilities of a whole organism, and so in order to assess the estrogenicity of the metabolic products, the parent phthalate would have to be artificially "metabolised" prior to testing in these assays. As far as we know, this procedure is not routinely applied. BBP, DBP, DIBP, and DINP also increased the proliferation rate of ZR-75 cells [97]. However, it must be noted that to date DINP has been reported to be estrogenic in this one study only, and not in any other. BBP, DBP, and DIBP also stimulated proliferation of MCF-7 cells, and this was one instance where BBP achieved a near-maximal dose response [97].

Zacharewski et al. [92] recently reported the results of testing 8 phthalates (DEHP, DBP, BBP, DHP, DIHP, DnOP, DINP, and DIDP) in a battery of *in vitro* and *in vivo* assays, and found DBP, BBP, and DHP to be weakly estrogenic in both an ER competitive binding assay and in a gene expression assay employing transfected MCF-7 cells. The response in the latter assay reached a maximum of 42% of the response to E2, when BBP was tested, with those of DBP and DHP reaching 36% and 20%, respectively. In transfected HeLa cells, only BBP stimulated an estrogenic response. The *in vivo* assays undertaken in this study included a uterine wet weight assay, and a vaginal cell cornification assay, the endpoints of which are controlled by estrogenic stimuli. Rats were dosed orally with high concentrations of phthalate (20, 200, and 2000 mg kg^{-1}). Despite BBP, DBP, and DHP displaying definitive interaction with the estrogen receptor *in vitro*, none of the phthalates tested were able to induce estrogenic responses in these *in vivo* experiments [92].

Blom et al. [98] reported that DEHP was estrogenically active, and as strongly so as 4-nonylphenol (4-NP, a chemical widely accepted to be a weak xenoestrogen), in an *in vitro* MCF-7 cell proliferation assay. This result was unusual in that, in previous reports where DEHP has been found to induce a response greater than that of the control, this has been extremely slight, so much so that it has been equivocal. Further, any phthalates which have been tested in any *in vitro* assay in which 4-NP has also been included have been far less potent than this "model" xenoestrogen [93, 95–97, 99]. It is possible that Blom et al.'s finding of appreciable estrogenicity of DEHP is a real result, but it is also conceivable that this response was due to some form of contamination of the chemical or equipment used.

Of all the phthalates, only BBP was tested in a novel assay proposed by Bolger et al. [99]. This is essentially a receptor binding assay, but it employs fluorescence polarisation rather than the traditional radioactive ligand, and is performed at room temperature as opposed to 4°C. In general, the data produced were similar to those derived using the traditional method, and BBP was found to be weakly interactive with recombinant ER-alpha (there are two estrogen receptors, namely ER-alpha and ER-beta; most research to date has been done using ER-alpha).

Knudsen and Pottinger [100] tested DEP, DAP, DBP, BBP, and DEHP in a rainbow trout estrogen receptor binding assay. All were active, although the responses were less than 30% of the maximum, and the response curves were not parallel to that produced by the positive control, E2. The same chemicals were tested for affinity for the rainbow trout testosterone and cortisol receptors, but none was found to interact with either of these receptors.

Very recently, structure-activity relationships of phthalates interacting with the estrogen receptor have been examined [101]. Alkyl phthalates with a chain length of up to 8 carbons were tested for their ability to displace radiolabelled E2 from a recombinant human estrogen receptor. BBP was not included in this study. Phthalates with a chain length of 3 and 4 carbons bound to the ER with a full response. Of the phthalates which tested positive, several isomeric forms were tested to assess structural requirements for interaction with the ER, along with different ring isomers of diallyl phthalate: namely *ortho* (1,2), *meta* (1,3), and *para* (1,4). The results revealed that a branched propyl chain was far less potent than a straight chain isomer, which the authors hypothesised was due to the size and bulk of the isomer. They concluded also that the *ortho*-isomer of diallyl phthalate (most of the commercial phthalates are of the *ortho*-form) was more potent than either the *meta*- or the *para*-isomer, but did not speculate as to the reason for this. It was suggested that a more hydrophobic structure (achieved by increasing the chain length from 3 to 4) evoked a stronger interaction with the receptor, but that a chain length of 4 carbons was the threshold for this behavioural pattern; above this, it was hypothesised that the alkyl group was too large and interfered with ligand-receptor interactions.

In summary, the relatively recent report of Jobling at al. [91] that some phthalates are weakly estrogenic in *in vitro* assays triggered a considerable amount of research. The results of this wave of research are still in the process of being published, but are, in general, very consistent. They show that some phthalates are very weakly active; even the most "potent" one (BBP) is about one million times less potent than E2.

3.2.2
In Vivo

The relatively recent realisation that some phthalates are estrogenic *in vitro* has obviously raised concerns that they may also be estrogenic *in vivo*. Data discussing the *in vivo* estrogenicity (or lack of it) of phthalates are scarce, but have recently begun to emerge. The reason for the slow appearance of such data is the length of time required to undertake sound *in vivo* studies, together with their expense, and the lag time prior to such data reaching the public domain. Therefore, from the first reported incidence of estrogenicity (*in vitro*) in 1995 [91], to date only a few articles have been published where the aim of the study was to assess estrogenic effects of phthalates *in vivo*. The following discussion includes some reports of effects which may not have been specifically estrogenically stimulated, however, the papers have been written with endocrine disruption in mind, as opposed to those which may contain descriptions of similar effects, but where the experiments were designed to assess general reproductive toxicity (these are discussed above in the section entitled "Reproductive Effects"). The division of such reports between sections is, however, at times arbitrary.

The first documented evidence of apparent estrogenicity of a phthalate *in vivo* was published by Sharpe et al. [102]. Wistar rats were dosed with BBP via drinking water, from 2 weeks prior to mating, through gestation, and until day 22 after birth. These authors observed a decrease in testis weight, along with a decrease in relative testis weight and testis/kidney weight ratio, and also a decrease in daily sperm production, which was proportional to the drop in testis weight. The hypothesis was put forward that the estrogenicity of BBP had suppressed production of FSH, which in turn had reduced the proliferation of

Sertoli cells in the testes of the developing fetuses. Since the ultimate size of the testis is controlled by the number of Sertoli cells, as is the number of germ cells which will go on to develop into spermatozoa, this could account for the effects observed in these animals. It was clearly stated by the authors that there was no evidence of a specifically estrogenic mechanism, although it is unlikely that it was a purely toxic one, since body weight and kidney weight of the exposed animals were normal.

Since the publication of these data, attempts have been made to repeat the experiment [103]. Minor changes were made to the design of the experiment. These included using a different strain of rats; the rats were not subject to premating exposure; the dose of BBP was probably slightly higher due to the use of glass water vessels rather than plastic ones; and the numbers of animals used per group were increased substantially. There were no adverse developmental effects observed in these rats. The reason for the different results obtained from the two studies is unclear, and hence Ashby et al. [103] were hesitant to refute Sharpe's study, and instead described their data as "failing to confirm" those reported by Sharpe et al. [102]. Sharpe et al. [104] have since commented on their original study. They can no longer observe the effects seen in their earlier study, but note the many unknown variables (such as variable organ weights both within and between strains of rats, and food consumption) that may have confounded their initial study.

There are some reports of phthalates having been tested in mammals *in vivo* where the endpoint was specifically dependent upon an estrogenic stimulus. Examples of such endpoints include an increase in relative uterine weight; vaginal cell cornification; and increased uterine vascular permeability. BBP was found to be inactive in the uterotrophic assay [96]. None of the phthalates tested by Milligan et al. [105] – namely DOP, DBP, BBP – stimulated an increase in uterine vascular permeability in ovarectimised mice. Zacharewski et al. [92] tested eight different phthalates (see above) in the uterine wet weight and the vaginal cell cornification assays. All were inactive as estrogens. Thus, there is a consensus of opinion presently that the phthalates tested to date do not demonstrate estrogenic activity, even when administered (orally) at high doses, in short-term rodent bioassays.

There are minimal data relating to the possible estrogenicity of phthalates in the aquatic environment. The production of the egg-yolk protein, vitellogenin, is an estrogen-dependent process, and therefore can be used as a biomarker for estrogen exposure of wild and aquarium fish [106]. This applies particularly to males, or immature females, where basal vitellogenin concentrations are extremely low. Christiansen et al. [107] injected high doses of DBP and BBP into immature rainbow trout as part of a wider, preliminary screen to identify compounds which should be investigated further. DBP did not induce vitellogenin production in any of the treated fish, whereas BBP induced a very weak (3-fold) increase in vitellogenin concentration. It is not entirely clear what relevance this route of exposure has to a real environmental situation, but this was purely a preliminary study to assess the activity of a chemical in a worst case scenario, presumably to be succeeded by a more realistic dosing regime. However, follow-up data have not yet been reported.

Harries et al. [108] exposed adult, breeding fathead minnows to 100 µg BBP L^{-1} via the water in a flow-through experimental system. BBP did not elicit an estrogenic response after a three-week exposure. Vitellogenin was measured in the plasma of these fish; the data can be seen in Fig. 4. Control fish were either exposed to untreated tank water, or to the carrier solvent (methanol) at the same concentration as were the treated fish. The fish dosed with 4-nonylphenol (4-NP) can in this case be considered as serving as a positive control, given that this chemical is widely accepted to be an estrogen-mimic. Harries et al. also determined the number of eggs produced by each spawning pair of fathead minnows. There was no significant difference in mean egg numbers produced pre-treatment compared to those produced when the same pairs of fish were exposed to BBP. Thus, BBP at 100 µg L^{-1} had no obvious endocrine-mediated effects in this experiment [108].

Knudsen et al. [109] monitored two alternative endpoints in juvenile rainbow trout injected with BBP at concentrations of 5 and 50 mg kg^{-1}. The estrogen-dependent endpoints in this experiment were the upregulation of the ER in the liver, and the induction of hepatic zona radiata proteins (Zrp). This latter protein is more sensitive to estrogens than is vitellogenin, according to Arukwe et al. [110]. BBP had no significant effect on the ER binding capacity at either concentration, and actually decreased Zrp levels. The reason for the decrease in Zrp levels is not known, but it was conjectured to be some kind of antagonistic effect at the receptor level. Whatever the explanation, there was no evidence of

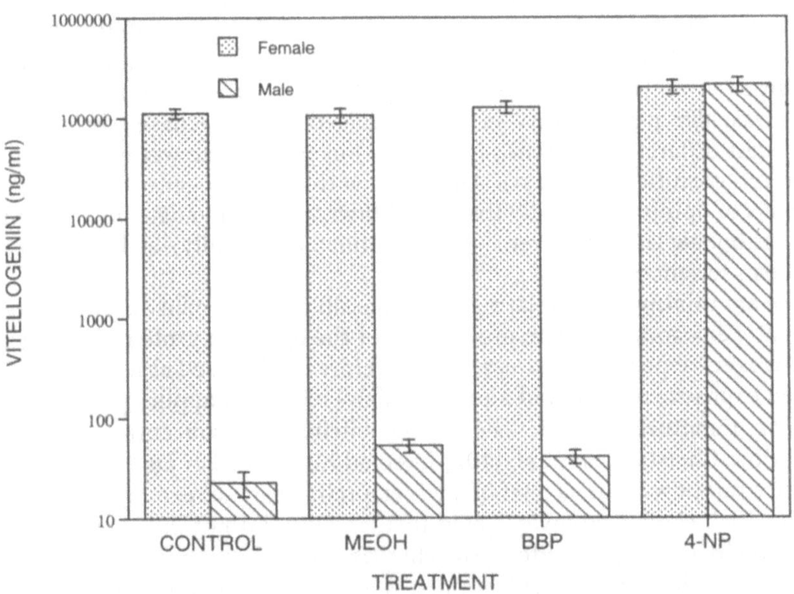

Fig. 4. Vitellogenin concentrations in blood plasma of male and female fathead minnows following a 3-week exposure to 100 µg 4-NP L^{-1}, 100 µg BBP L^{-1}, or carrier solvent (MEOH). Note that whereas 4-NP demonstrated estrogenic activity (which was particularly noticeable in males), BBP showed no estrogenic effect. Adapted from Harries et al. [108]

any estrogenic activity of BBP. Thus, although there are very little data available on which to base a conclusion, it seems unlikely that phthalates will demonstrate any significant estrogenic effects *in vivo* in fish, especially at environmentally-relevant concentrations. Nevertheless, there is a need for a few multi-generation studies, using at least one fish species and one invertebrate species, and a variety of concentrations of the test phthalate, before robust conclusions can be drawn.

3.3 Anti-Androgenic Activity of Phthalates

The most recent hypothesis on the mechanisms behind the reproductive toxicity of phthalates is that, rather than behaving as estrogen mimics, they may in fact be having an inhibitory effect on androgen activity; that is, that they are 'anti-androgens'. The first documented evidence for this idea was published in 1998, when Sohoni and Sumpter [111] reported BBP to have antagonistic activity in a recombinant yeast-based androgen screen. Subsequent to these data being published, several more phthalates, plus some of their metabolites, were assessed for anti-androgenic activity using the method described in Sohoni and Sumpter [111]. Some of the data obtained from these experiments are shown in Fig. 5 (our unpublished data). Figure 5 shows that the major metabolites of

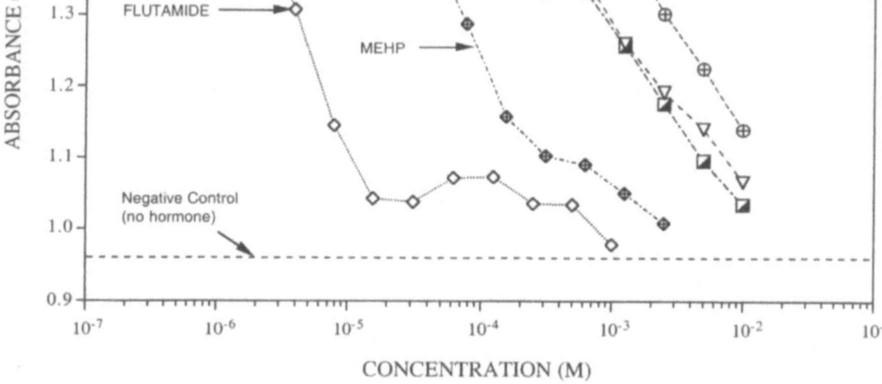

Fig. 5. The anti-androgenic activity of BBP, MBP, MBzP, and MEHP. Each chemical was serially diluted from 10^{-2} M, and compared to the activity of the clinical anti-androgen flutamide, which was serially diluted from 10^{-3} M. Anti-androgenic activity was detected using a recombinant yeast-based assay (as described by Sohoni and Sumpter [111]). The figure illustrates the ability of these chemicals to block the normal androgenic response to DHT

DEHP (MEHP), DBP (MBP), and BBP (MBP and MBzP) can block the binding of dihydrotestosterone (DHT) to the androgen receptor. The metabolites were found to be more potent in this assay than their respective parent phthalates. This is potentially highly significant data, when it is considered that the weak estrogenic activity of phthalates is often thought of as not constituting a risk, due to their rapid metabolism to the estrogenically inactive monoesters *in vivo*, and also in the environment.

Also in 1998, results of an *in vivo* study, describing the behaviour of DBP as similar to that of the clinical anti-androgen, flutamide, was published [112]. In this study, Sprague-Dawley rats were orally administered high doses (250, 500, and 750 mg kg^{-1} day^{-1}) throughout gestation and lactation, until postnatal day 20. The spectrum of adverse reproductive effects in the male offspring included decreased anogenital distance, malformed epididymides, testicular atrophy, hypospadias, ectopic or absent testes, absent prostate gland and seminal vesicles, and diminished size of testes and seminal vesicles. In contrast, estrogen-dependent endpoints in female offspring, such as vaginal opening and estrous cyclicity, were not affected. Fig. 6 illustrates data from some of the parameters measured in this study.

A subsequent study by the same authors [113] covered a shorter exposure period which encompassed gestation days 12 to 21; that is, the period during

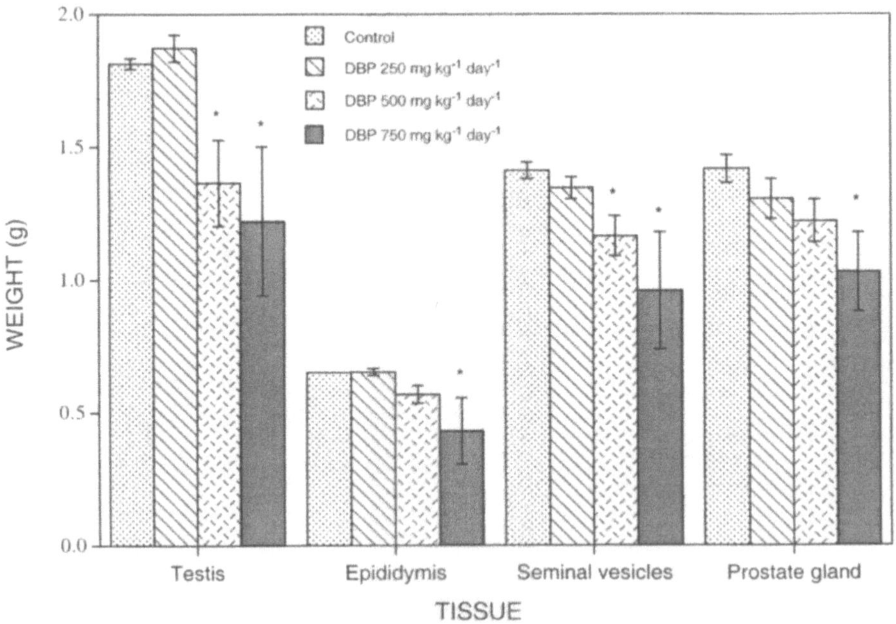

Fig. 6. *In vivo* anti-androgenic effects of DBP. The data were taken from Mylchreest et al. [112], and show adverse effects of 250, 500, and 750 mg DBP kg^{-1} day^{-1} on androgen regulated endpoints in rats. The DBP was given to pregnant rats throughout gestation and lactation, through to postnatal day 20, and organ weights recorded from the male offspring. * indicates a significant difference from the control group ($p < 0.05$)

which androgen-dependent prenatal male reproductive tract development occurs. Despite widespread adverse reproductive effects in offspring of DBP treated rats, many of which were common to both DBP and flutamide, there were some which were distinct to either one or the other of the chemicals. These included a high incidence of prostate agenesis in flutamide-exposed offspring compared to the incidence found in DBP treated rats, and the blocking of the inguinoscrotal descent caused by DBP. These authors concluded that exposure of fetuses to DBP during the sensitive window of gestation (days 12–21, covering the period of sexual differentiation in the rat), can lead to severe reproductive disruption in male offspring. Although the effects seen were considered to be mediated via an androgen-regulated pathway, it was hypothesised that the mechanism did not involve direct interaction with the androgen receptor (AR), since the behaviour of DBP differed from that of flutamide in some ways. This interpretation was supported by the results of Gray et al. [114], who found DBP and MBP not to have an agonistic interaction with the AR in receptor binding and transcriptional activation assays. However, the data shown in Fig. 5 suggests that MBP – the metabolite of DBP likely to be found in the exposed rats – does have a receptor mediated antagonistic mechanism.

These results, demonstrating an anti-androgenic action of DBP *in vivo*, are very important for a number of reasons. Firstly, they demonstrate that, despite several decades of research on DBP, unexpected, and important, effects can still be discovered. Secondly, that despite the failure to observe estrogenic effects in females (the research was undertaken originally to see if the estrogenic effects observed *in vitro* were also manifest *in vivo*), effects indicative of another type of endocrine activity, namely anti-androgenicity, were observed in males. Thirdly, that the timing of exposure is all important; exposure produces effects only if it occurs during the relatively brief period of sexual differentiation. Subsequent research can now involve lower doses, to determine if there is a real risk at realistic levels of exposure. Such research can obviously only be conducted using *in vivo* experimental designs.

4 Discussion

In this chapter, we have attempted to evaluate the extent of exposure of organisms to phthalates, along with their activity as endocrine disrupting chemicals, thus giving an overall picture of the likelihood that this class of chemicals are capable of causing adverse reproductive effects in either aquatic or terrestrial organisms.

Both parameters are almost impossible to quantify presently. Exposure does not take place via a single point source, but through various diffuse sources, and does not occur along a specific (e.g., oral) route, but potentially through several, most of which have not been thoroughly evaluated, if evaluated at all. Exposure of aquatic organisms likewise does not take place simply through uptake and absorption from water, but also as a result of intake of suspended and bed sediments, and of food sources. Hence, the issue of exposure, as with many other abundant environmental contaminants, is extremely complex. We have attempted to illustrate the possible concentrations of phthalates in given environments, and in products to which humans might be exposed, but we cannot calculate precise exposure rates. One of the more easily evaluated routes is that occurring via food eaten by humans: this has been undertaken by MAFF [45], who estimated a total phthalate intake via this route to be up to 1.6 mg $person^{-1}$ day^{-1} (for a high level intake). This would, however, be a maximal intake for an adult of 60 kg weight. A child, or a young toddler, who would be more sensitive to developmental toxicants, would most likely have a far lower intake from the same sources. Until considerably more is known about exposure to phthalates, it is very difficult to conduct meaningful biological studies on which risk assessment can be based.

Once the phthalates have been absorbed into the organism, there is a high probability that they will be rapidly metabolised, firstly into their monoesters, and then into various degradation products, until they are broken down completely into carbon dioxide and water. The shorter the alkyl chain of an individual phthalate, the more rapidly it will be metabolised. There is not enough information on degradation of phthalates following dermal exposure or inhalation to be able to state definitively whether the metabolic process follows the same pattern and rate as that occurring after oral exposure. It has been suggested that the lack of metabolism of DEHP observed in some experiments is due to saturation following extremely high doses [56], and it seems likely that realistic exposures would not be sufficient to reach saturation. This leads us back to the issue of exposure estimates once more. Also, it is not clear whether the presence of other organic contaminants in the blood stream would lower this saturation point with regard to phthalates, which again is an extremely complex question to address realistically.

With respect, specifically, to their endocrine disrupting potential, it is clear that some, but not all, of the phthalates (the most potent being BBP, DBP, and DIBP) are capable of binding to the estrogen receptor. However, this activity is not observed in *in vivo* assays; no estrogenic effects of any phthalate have yet been demonstrated *in vivo*, although it must be said that this statement is based on relatively few studies to date. However, recent studies suggest that at least one phthalate, DBP, has anti-androgenic effects in rodents, albeit at high doses. Even less studies in which phthalates have been investigated for endocrine activity in non-mammalian vertebrates, or invertebrates, have been reported. This is despite the fact that phthalates are ubiquitous in the environment, and hence most, if not all, wildlife will receive exposure, albeit to varying degrees. In the aquatic environment, it would be wise to focus activity in particular on organisms that live on, or in, the sediment, where the highest concentrations of phthalates are found. Invertebrates as well as vertebrates should be studied. Most of this research will need to be laboratory-based, because the environment is polluted with an extremely diverse cocktail of anthropogenic chemicals, and hence separating out any effects due to phthalates from those due to other chemicals (or mixtures of chemicals) will be extremely difficult, if not impossible. Once appropriate tests, incorporating a number of endocrine-related endpoints, have been established and validated (a programme to do this is underway presently), then phthalates, because of their widespread and major use, should be priority chemicals for testing.

Despite phthalates not eliciting an estrogen-specific response in mammalian species *in vivo*, it is nonetheless clear that they are capable of disrupting the reproductive development *in utero* of young mammals [79, 80, 82, 83]. The question therefore arises, "what are the mechanisms behind this activity?" Is it possible that we have been looking in the wrong direction, and that what is actually happening is that the phthalates are having an inhibitory effect on the androgen receptor? This is without doubt an area which needs to be investigated in greater depth, particularly given that the majority of adverse effects observed following exposure of mammalian species, *in vivo*, occur in the male offspring. As well as investigating this issue further in mammalian species, there is a need to monitor specific anti-androgenic endpoints in aquatic organisms. However, presently there is a paucity of suitable endpoints, especially in lower vertebrates and invertebrates. Thus, for example, despite having an excellent indicator of estrogen action in oviparous vertebrates such as fish and amphibians (vitellogenin), there are presently no indicators of either androgenic or anti-androgenic activity. Thus, even if phthalates (or any other chemicals) did have androgenic and/or anti-androgenic activity *in vivo*, this would be very difficult to detect in many species, especially those, such as fish, used in aquatic toxicity tests.

Overall, the concentrations at which adverse effects have been reported to occur have been far higher than those which we would expect organisms to be exposed to in a real environment. It seems unlikely that the low levels of phthalates to which humans are exposed are capable of causing the trends in adverse reproductive health of males over the past few decades, such as decreased sperm counts (if true), and increased cases of testicular cancer, cryptorchidism, and hypospadias. But it is theoretically possible that they may be a contributory factor. The biggest problem in this area is that humans, and wildlife, are exposed simultaneously to complex mixtures of chemicals, which might (or might not) have some net adverse effect, whereas almost all toxicology is based upon the testing of one chemical on a very limited range of species. These chemicals could interact (to produce an effect) in various ways, ranging from one antagonising another, thereby reducing the anticipated effect, through to one synergising with another, thereby increasing the anticipated effect. The phenomenon of synergism, though widely speculated upon, has not yet been proven to occur, either *in vitro* or *in vivo*, but it is accepted that environmental estrogens at least can have additive activity when assayed together. For example, Jobling et al. [91] found BBP and DBP to have agonistic effects when in the presence of E2 in a transcriptional activation assay. This complicated, but very important, area of research is only just beginning.

What has become clear is that, despite large and widespread use of phthalates for around 50 years, and extensive research on their acute and chronic effects on a range of organisms, there is still a great deal to learn about this group of chemicals. This is amply demonstrated by the present ongoing research on the endocrine activities of phthalates. None of the estrogenic (*in vitro*) and anti-androgenic (*in vivo*) activities of some phthalates were expected or predicted; a finding of only 4 years ago [91] opened up this whole new area of research on phthalates, which is now expanding rapidly, and obviously has a long way still to go. It is quite likely that other unexpected findings will emerge from this intensified activity on the physiological effects of phthalates. What are required now are experiments which incorporate realistic and chronic exposure regimes in, preferably, multi-generation studies, or if not, studies in which the particularly sensitive stages of the life cycle are exposed. Varying routes of exposure, and a range of test organisms (from invertebrates to rodents) should be used in these studies. Only when such studies are completed will it be possible to assess the risk to humans and wildlife due to exposure to phthalates.

5 References

1. Giam CS, Atlas E, Powers MA, Leonard JE Jr. (1984) Phthalic acid esters. In: Hutzinger O (ed), Handbook of Environmental Chemistry. Springer, Berlin Heidelberg, pp 67–142
2. Lyons G (1995) WWF Report (25/7/95), World Wildlife Fund (WWF) UK, Godalming, Surrey
3. Stringer R, Labounskaia I, Santillo D, Johnston P, Siddorn J, Stephenson A (1997) Greenpeace Research Laboratories Technical Note 06/97. University of Exeter, UK
4. Rastogi SC (1998) Chromatographia 47:724
5. Sugatt RH, O'Grady DP, Banerjee S, Howard PH, Gledhill WE (1984) Appl Environ Microbiol 47:601
6. Staples CA, Peterson DR, Parkerton TF, Adams WJ (1997) Chemosphere 35:667
7. Thuren A (1986) Bull Environ Contam Toxicol 36:33
8. van der Velde EG, Belfroid AC, de Korte GAL, van der Horst A, Versteegh AFM, Schafer AJ, Vethaak AD, Rijs GBJ (1999) Poster Presentation; Eidgenössische Anstalt für Wasserversorgung, Abwasserreinigung und Gewässerschutz (EAWAG), Monte Verita, Ascona, Switzerland 8th-12th March
9. Long JLA, House WA, Parker A, Rae JE (1998) Sci Tot Environ 210/211:229
10. Giam CS, Chan HS, Neff GS, Atlas EL (1978) Science 199:419
11. Law RJ, Fileman TW, Matthiessen P (1991) Wat Sci Technol 24:127
12. Vitali M, Guidotti M, Macilenti G, Cremisini C (1997) Environ International 23:337
13. Tan GH (1995) Bull Environ Contam Toxicol 54:171
14. Fatoki OS, Vernon F (1990) Sci Tot Environ 95:227
15. Fatoki OS, Ogunfowokan AO (1993) Environ International 19:619
16. Sheldon LS, Hites RA (1978) Environ Sci Technol 12:1188
17. Pirie D, Steven L, McGrory S, Best G (1996) Scottish Environmental Protection Agency Report (August 1996), SEPA, Stirling, Scotland
18. Thuren A, Larsson P (1990) Environ Sci Technol 24:554
19. Giam CS, Atlas E, Chan HS, Neff GS (1980) Atmos Environ 14:65
20. Kirchmann H, Tengsved A (1991) Swedish J Agri Res 21:115
21. Shelton DR, Boyd SA, Tiedje JM (1984) Environ Sci Technol 18:93
22. Ejlertsson J, Meyerson U, Svensson BH (1996) Biodegradation 7:345
23. Ejlertsson J, Alnervik M, Jonsson S, Svensson BH (1997) Environ Sci Technol 31:2761
24. O'Grady DP, Howard PH, Werner AF (1985) Appl Environ Microbiol 49:443
25. Scholz N, Diefenbach R, Rademacher I, Linnemann D (1997) Bull Environ Contam Toxicol 58:527
26. Jianlong W, Ping L, Hanchang S, Yi Q (1997) Chemosphere 35:1747
27. Roslev P, Madsen PL, Thyme JB, Henriksen K (1998) Appl Environ Microbiol 64:4711
28. Group EF Jr. (1986) Env Health Perspect 65:337

29. Saeger VW, Tucker ES (1976) Appl Environ Microbiol 31:29
30. Adams WJ, Saeger VW (1993) Utility of microcosms for predicting the environmental fate of chemicals: a comparison of two microcosm designs with butyl benzyl phthalate. In: Gorsuch JW, Dwyer FJ, Ingersol CG, LaPoint TW (eds), Environmental Toxicology and Risk Assessment: Aquatic, Plant and Terrestrial:2nd Volume. American Society for Testing and Materials (ASTM), Standards and Technical Publications (STP) 1216, Philadelphia
31. Furtmann RNK (1996) European Chemical Industry Council (CEFIC) Report 6/93 (Brussels)
32. Rogers HR (1996) Sci Tot Environ 185:3
33. Jaeger RJ, Rubin RJ (1972) New England J Med 287:1114
34. Lygre H, Solheim E, Gjerdet NR, Berg E (1993) Acta Odontol Scand 51:45
35. Steiner I, Scharf L, Fiala F, Washuttl J (1998) Food Add Contam 15:812
36. Vinkelsoe J, Jensen GH, Johansen E, Carlsen I, Rastogi SC (1997) Danish Environmental Protection Agency 15.4.97
37. Environmental Data Services (1996) ENDS Report 257:25
38. Environmental Data Services (1997) ENDS Report 268:30
39. Environmental Data Services (1998) ENDS Report 286:25
40. US Consumer Product Safety Commission (CPSC) (1998) News Release 99-031
41. UK Ministry of Agriculture, Fisheries, and Food (MAFF) (1995) Food Surveillance Information Sheet No 60
42. van Lierop JBH (1997) Food Add Contam 14:555
43. Nerin C, Salafranca J, Rubio C, Cacho J (1998) Food Add Contam 15:842
44. Harrison N (1998) Food Add Contam 5:493
45. UK Ministry of Agriculture, Fisheries, and Food (MAFF) (1996) Food Surveillance Information Sheet No 82
46. UK Ministry of Agriculture, Fisheries, and Food (MAFF) (1996) Food Surveillance Information Sheet No 83
47. UK Ministry of Agriculture, Fisheries, and Food (MAFF) (1998) Food Surveillance Information Sheet No 168
48. Sharman M, Read WA, Castle L, Gilbert J (1994) Food Add Contam 11:375
49. Petersen JH (1991) Food Add Contam 8:701
50. Page BD, Lacroix GM (1995) Food Add Contam 12:129
51. Environmental Data Services (1995) ENDS Report 245:7
52. Dirven HAAM, van den Broek PHH, Jongeneelen FJ (1993) Int Arch Occup Environ Health 64:555
53. Dirven HAAM, van den Broek PHH, Arends T, Noordkamp EM, de Lepper AJTM, Henderson PT, Jongeneelen FJ (1993) Int Arch Occup Environ Health 64:549
54. Ng KME, Chu I, Bronaugh RL, Franklin CA, Somers DA (1992) Toxicol Appl Pharmacol 115:216
55. Oishi S, Hiraga K (1982) Arch Toxicol 15:149
56. Schulz CO, Rubin RJ (1973) Environ Health Perspect 3:123
57. Saillenfait AM, Payan JP, Fabry JP, Beydon D, Langonne I, Gallissot F, Sabate JP (1998) Toxicol Sci 45:212
58. Coldham NG, Dave M, Sauer MJ (1998) J Mass Spectrom 33:803
59. Wofford HW, Wilsey CD, Neff GS, Giam GS, Neff GM (1981) Ecotoxicol Environ Saf 5:202
60. Yan H, Ye C, Yin C (1995) Env Toxicol Chem 14:931
61. Stalling DL, Hogan JW, Johnson JL (1973) Environ Health Perspect 3:153
62. Barron MG, Albro PW, Hayton WL (1995) Env Toxicol Chem 14:873
63. Carr KH, Coyle GT, Kimerle RA (1997) Env Toxicol Chem 16:2200
64. Gangolli SD (1982) Environ Health Perspect 45:77
65. Gray TJB, Gangolli SD (1986) Environ Health Perspect 65:229
66. Hardell L, Ohlson C-G, Fredrikson M (1997) Int J Cancer 73:828
67. Jones HB, Garside DA, Liu R, Roberts JC (1993) Exp Molec Pathol 58:179

68. Richburg JH, Boekelheide K (1996) Toxicol Appl Pharmacol 137:42
69. Li L-H, Jester WF, Orth JM (1998) Toxicol Appl Pharmacol 153:258
70. Fredricsson B, Moller L, Pousette A, Westerholm R (1993) Pharmacol Toxicol 72:128
71. Pennanen S, Tuovinen K, Huuskonen H, Kosma V-M, Komulainen H (1993) Fund Appl Toxicol 21:204
72. Siddiqui A, Srivastava SP (1992) Bull Environ Contam Toxicol 48:115
73. Murature DA, Tang SY, Steinhardt G, Dougherty RC (1987) Biomed Environ Mass Spectrom 14:473
74. Davis BJ, Maronpot RR, Heindel JJ (1994) Toxicol Appl Pharmacol 128:216
75. Laskey JW, Berman E (1993) Reprod Toxicol 7:25
76. Ema M, Kurosaka R, Amaro H, Ogawa Y (1994) Reprod Toxicol 8:231
77. Ema M, Harazono A, Miyawaki E, Ogawa Y (1997) Bull Environ Contam Toxicol 58:636
78. Ema M, Kurosaka R, Amano H, Ogawa Y (1995) Toxicol Letters 78:101
79. Ema M, Miyawaki E, Kawashima K (1998) Toxicol Letters 98:87
80. Imajima T, Shono T, Zakaria O, Suita S (1997) J Ped Surgery 32:18
81. Agarwal DK, Maronpot RR, Lamb JC, Kluwe WM (1985) Toxicology 35:189
82. Piersma AH, Verhoef A, Dortant PM (1995) Toxicology 99:191
83. Wine RN, Li L-H, Hommel Barnes L, Gulati DK, Chapin RE (1997) Environ Health Perspect 105:102
84. Sharpe RM (1998) Pure Appl Chem 70:1685
85. DeFoe DL, Holcombe GW, Hammermeister DE, Biesinger KE (1990) Env Toxicol Chem 9:623
86. McCarthy JF, Whitmore DK (1985) Env Toxicol Chem 4:167
87. Rhodes JE, Adams WJ, Biddinger GR, Robillard KA, Gorsuch JW (1995) Env Toxicol Chem 14:1967
88. Abernathy SG, Mackay D, McCarthy LS (1988) Environ Toxicol Chem 8:163
89. Davis WP (1988) Environ Biol Fishes 21:81
90. Staples CA, Adams WJ, Parkerton TF, Gorsuch JW, Biddinger GR, Reinert KH (1997) Env Toxicol Chem 16:875
91. Jobling S, Reynolds T, White R, Parker MG, Sumpter JP (1995) Environ Health Perspect 103:582
92. Zacharewski TR, Meek MD, Clemons Jd, Wu ZF, Fielden MR, Matthews JB (1998) Toxicol Sci 46:282
93. Soto AM, Sonnenschein C, Chung KL, Fernandez MF, Olea N, Olea Serrano F (1995) Environ Health Perspect 103 (Suppl 7):113
94. Gaido KW, Leonard LS, Lovell S, Gould JC, Babai D, Portier CJ, McDonnell DP (1997) Toxicol Appl Pharmacol 143:205
95. Petit F, LeGoff P, Cravedi J-P, Valotaire Y, Pakdel F (1997) J Molec Endocrinol 19:321
96. Coldham NG, Dave M, Sivapathasundaram S, McDonnell DP, Connor C, Sauer MJ (1997) Environ Health Perspect 105:734
97. Harris CA, Henttu P, Parker MG, Sumpter JP (1997) Environ Health Perspect 105:802
98. Blom A, Ekman E, Johannisson A, Norrgen L, Pesonen M (1998) Arch Environ Contam Toxicol 34:306
99. Bolger R, Wiese TE, Ervin K, Nestich S, Checovich W (1998) Environ Health Perspect 106:551
100. Knudsen FR, Pottinger TG (1999) Aquat Toxicol 44:159
101. Nakai M, Tabira Y, Asai D, Yakabe Y, Shimyozu T, Noguchi M, Takatsuki M, Shimohigashi Y (1999) Biochem Biophys Res Commun 254:311
102. Sharpe RM, Fisher JS, Millar MM, Jobling S, Sumpter JP (1995) Environ Health Perspect 103:1136
103. Ashby J, Tinwell H, Lefevre PA, Odum J, Paton D, Millward SW, Tittensor S, Brooks AN (1997) Regul Toxicol Pharmacol 26:102
104. Sharpe RM, Turner KJ, Sumpter JP (1998) Environ Health Perspect 106:A220
105. Milligan SR, Balasubramanian AV, Kalita JC (1998) Environ Health Perspect 106:23
106. Sumpter JP, Jobling S (1995) Environ Health Perspect 103 (Suppl 7):173

107. Christiansen LB, Pedersen KL, Korsgaard B, Bjerregaard P (1998) Marine Environ Res 46:137
108. Harries JE, Runnalls T, Hill E, Harris CA, Maddix S, Sumpter JP, Tyler CR (2000) Env Sci Technol 34:3003
109. Knudsen FR, Arukwe A, Pottinger TG (1998) Environ Pollution 103:75
110. Arukwe A, Knudsen FR, Goksoyr A (1997) Environ Health Perspect 105:1
111. Sohoni P, Sumpter JP (1998) J Endocrinol 158:327
112. Mylchreest E, Cattley RC, Foster PMD (1998) Toxicol Sci 43:47
113. Mylchreest E, Sar M, Cattley RC, Foster PMD (1999) Toxicol Appl Pharmacol 156:81
114. Gray LE, Ostby JS, Mylchreest E, Foster PMD, Kelce WR (1998) Toxicol Sci 42:176

Subject Index

A
activator function (AF)
–, AF-1 5, 10, 17
–, AF-2 9–11, 16–17
activator protein 1 (AP-1) 18, 19
additive effects 43, 49–53
AF, see activator function
albumin 7, 140
alkylphenol polyethoxylates 75, 129, 133–137
alkylphenols 75, 129, 132, 137–139
alligators 41, 132
amsonic acid 79
androgen 131–132, 149
–, action 43
–, receptor 41–42, 51, 54, 164
–, response element 46
androstenedione 7
anogenital distance 47–48
antiandrogens 41, 46, 49, 51, 164
antiestrogenicity 15
antiestrogens 12, 14, 17–20
–, triphenylethylene-type 72
–, HO-PCBs 159, 161
antihormones 50–53
antioxidant 102, 114, 120
antithyroid 58
AP-1, see activator protein 1
apigenin 69
aromatase 86, 94
arteriosclerosis 95
assays
–, for androgen receptor agonists and antagonists 54
–, cell proliferation assay 33–34
–, competitive binding assay 28–29
–, DNA binding and transcriptional activation assay 55
–, Hershberger assay 42, 58
–, mammalian-based reporter gene assay 31–33
–, N/C interaction assay 55
–, phage display assay 34
–, yeast-based steroid receptor assay 30–31
astringin 69
Atlantic salmon 139

B
BADGE, see BPA diglycidylether
BBP, see butylbenzylphthalate
bakelite 139, 149
behavior 131
benzylbutylphthalate, see butylbenzylphthalate
biochanin A
–, chemical structure 67, 104
–, concentration in animal feed and silage 109
–, exposure of domestic animals 108
–, metabolism 104–106
birds 132
2,2-bis(p-chlorophenyl)-1,1-dichloroethane (p,p′-DDT) 41, 44, 46, 48, 77, 131–132
2,2-bis(p-chlorophenyl)-1,1-dichloroethylene (p,p′-DDE) 41, 47–48, 58, 131, 156
–, binding to androgen receptor 46, 164
–, chemical structure 44, 77
–, environmental levels 165
–, inhibition of AR DNA binding 46
–, inhibition of AR N/C interaction 57
2,2-bis(p-hydroxyphenyl)-1,1,1-trichloroethane (HPTE) 41, 48–49, 52
–, binding to androgen receptor 46
–, chemical structure 44, 77
–, inhibition of AR DNA binding 46
–, inhibition of AR N/C interaction 57
bisphenol A (BPA)
–, chemical structure 75
–, derivatives 141–144
–, diglycidyl ether (BADGE) 75, 141–142, 144

–, diglycerolmethacryate (Bis-GMA) 141–143
–, dimethacrylate (BPA-DMA) 75, 142
–, inhibition of AR N/C interaction 57
–, triethyleneglycol dimethacryate (TEDGMA) 141–142
–, in utero exposure 145–146, 148
bone 142–143
bone mass 90, 95
BPA, see bisphenol A
BPA diglycidylether (BADGE)
–, chemical structure 75
–, occurrence 141–142
–, pharmacokinetics in mice 144
BPA-dimethacrylate (BPA-DMA)
–, chemical structure 75
–, occurrence 142
BPA-DMA, see BPA-dimethacrylate
breast cancer 89, 164
butylbenzylphthalate (BBP) 57, 78
butylphenol 57

C

Cape Cod, Massachusetts 135–136, 143
cardiovascular system 94
carp 138-139
catechin 69
cell lines
–, CHO 57
–, COS 54
–, CV1 55
–, HepG2 32–33, 158
–, LNCaP 33
–, MCF-7 33–34, 48, 187, 189, 158
–, MDA-MB-231 158
–, ZR-75 187-188
cell proliferation 131–132, 139, 140–142, 147, 149
cell proliferation assay 33–34
chaperone 8
chemical mixtures 135, 149
chlordecone (Kepone)
–, chemical structure 77
–, estrogenic activity 163
2-(o-chlorophenyl)-2-(p-chlorophenyl)-1,1,1-trichloroethane (o,p′-DDT) 131–132
–, chemical structure 77
–, estrogenic activity 163
–, inhibition of AR N/C interaction 57
CHO cells 57
clomiphene 72
co-activator 10–11, 16–17, 20, 51
cognitive function 95
competitive binding assay 28–29
co-repressor 10, 20, 51
corpus luteum 90–91
corticosteroid-binding globulin 140
COS cells 54
coumestans
–, chemical structure 68
–, as a phytoestrogen class 103, 106
coumestrol
–, biological effects 115–117, 123–124
–, chemical structure 68, 104
–, concentration in plant foods 106
–, occurrence 104
CV1 cells 55
cyproterone acetate
–, antiandrogenic activity 57
–, chemical structure 44, 73

D

daidzein
–, in breast milk 113
–, and cardiovascular disease 121
–, chemical structure 67, 104
–, concentration in animal feed and silage 109
–, exposure of domestic animals 108
–, occurrence and metabolism 104–106
DBD, see DNA binding domain
o,p′-DDT, see 2-(o-chlorophenyl)- 2-(p-chlorophenyl)-1,1,1-trichloroethane
p,p′-DDE, see 2,2-bis(p-chlorophenyl)-1,1-dichloroethylene
p,p′-DDT, see 2,2-bis(p-chlorophenyl)-1,1-dichloroethane
degradation
–, aerobic 135–137
–, anaerobic 135, 137
dehydroepiandrosterone sulfate 91
demographic differences 164
DES, see diethylstilbestrol
O-desmethylangolensin 67
detergents 134, 148
DHT, see 5α-dihydrotestosterone
di(ethylhexyl)phthalate (DEHP) 78
5α-dihydrotestosterone (DHT) 43–45
dibromochloropropane (DBCP) 79
DNA binding and transcriptional activation assay 55
DNA binding domain (DBD) 4–5, 9, 11
dibutylphthalate 57–58
dicarboximide fungicides 47
dieldrin
–, chemical structure 77
–, activity in E-screen assay 164

diethylstilbestrol (DES) 1, 3, 11–12, 15–16, 18
–, chemical structure 72, 103
–, structural similarity with resveratrol 106
–, in utero exposure 132, 146
dienestrol 72
dietary estrogens 111
5α-dihydrotestosterone (DHT)
–, chemical structures 71
–, transcriptional activation 52, 55
drinking water wells 134–135, 143

E
EAC, see endocrine active compounds
ectopic testis 47
EDSTAC, see Endocrine Disruptor Screening and Testing Advisory Committee
EGF, see epidermal growth factor
endocrine active compounds 1, 12, 14–15, 18–20
Endocrine Disruptor Screening and Testing Advisory Committee (EDSTAC) 41, 58–59
endometrium 132, 139
–, cancer 89
–, hyperplasia 89
endosulfan
–, activity in E-screen assay 164
–, chemical structure 77
enterodiol
–, chemical structure 68
–, formation from plant lignans 103, 105–106
–, production from plant food 105
enterolactone
–, chemical structure 68
–, formation from plant lignans 103, 105–106
–, production from plant food 105
Environmental Protection Agency (EPA) 41, 49, 54, 58–59
epicatechin 69
epidermal growth factor (EGF) 3, 11–12
epididymis 45, 139, 146
epiphyseal closure 87, 90
epoxy resin 141, 143
equilenin 65
equilin 65
equol
–, antioxidant effects 121
–, chemical structure 67
–, formation from daidzein 104, 106
ER, see estrogen receptor
ERE, see estrogen response element
ERKO, see estrogen receptor knockout
E-screen 132, 140, 164
estradiol 1, 7, 18, 83, 117, 119
17β-estradiol 7, 155
–, chemical structure 65, 103
–, importance for development and health 114
estriol 65, 83
estrogen 1, 3, 7, 9, 12, 17–20
–, estrous cyclicity 131, 146
–, estrus 146–147
–, excretion 84
–, gynecomastia 131
–, reference values 96
–, replacement therapy 95
–, resistance 86
–, stilbene-type 71
–, synthesis
–, – in adipose tissue 88
–, – in ovary 88
–, – in placenta 91
–, synthesis of transport proteins 92
–, target organs 85
–, ultrasensitive assay 96
–, uterine weight 142, 145, 147
–, vaginal cornification 147
–, vaginal opening 146
estrogen binding proteins
–, α-fetoprotein 140
–, albumin 7, 140
–, corticosteroid-binding globulin 140
–, sex hormone-binding globulin 84, 140
estrogen receptor (ER) 42, 139–140, 157
–, alpha (ERα) 85–86, 131, 161
–, beta (ERβ) 85–86, 131, 161
–, knockout mice 85–86
estrogen receptor knockout (ERKO)
–, αERKO 7
–, βERKO 7
estrogen response element (ERE) 9–11, 17–18, 140
–, consensus ERE (cERE) 9, 11, 17, 20
–, nonconsensus ERE (nERE) 11, 17, 20
estrogenicity 15, 155
estrone 65, 83
17α-ethinylestradiol 72
ethinyltestosterone 74
ethisterone, see ethinyltestosterone
European Committee for Food 142

F
female genital tract 131
–, endometrium 132, 139

–, ovary 131, 139
–, uterus 146–147, 149
–, vagina 132, 146–147, 149
feminization 132
fenarimol 79
fetal adrenal gland 92
fetal development 45, 163
fetal positioning effects 131, 145–146
α-fetoprotein 140
fish
–, Atlantic salmon 139
–, carp 138–139
–, feminization 132
–, Japanese medaka 139
–, mosquitofish 57
–, rainbow trout 137–138
flavonoids
–, chemical structures 69
–, estrogenic activity 106
flaxseed
–, biological effects 115–118, 122–124
–, as source of lignans 103, 105, 111
flutamide 73
follicle stimulating hormone (FSH) 146
follicular phase 90
Food Quality Protection Act 41
formononetin
–, chemical structure 67, 104
–, concentration in plant foods 105–106
–, in livestock feed and silage 108–109
fluorescence polarization 29
FSH, see follicle stimulating hormone
fungicide 46
Fusarium spp. toxins 101, 107, 109

G
galactorrhea 93
genistein 14, 18
–, biological effects 115–121, 123–124
–, in breast milk 113
–, chemical structure 67, 104
–, concentration in animal feed and silage 109
–, exposure of domestic animals 108
–, intake in Japan and USA 112
–, occurrence and metabolism 104–106
glycitein 67
gonadotropin releasing hormone 87
groundwater 137
growth 87
guaiaretic acid 68
gynecomastia 88, 90

H
heat-shock protein (hsp) 8
HepG2 cells 32–33, 158
2,2′, 3,4′, 5,5′, 6-heptachloro-4-biphenylol 163
Hershberger assay 42, 58
hexestrol 72
HO-PCB3, see 2′, 4′, 5′-trichloro-4-biphenylol
HO-PCB4, see 2′, 3′, 4′, 5′-tetrachloro-4-biphenylol
HPTE, see 2,2-bis(p-hydroxyphenyl)-1,1,1-trichloroethane
hsp, see heat-shock protein
16α-hydroxyestrone 65
hydroxyflutamide 44, 47–48, 56–57
hydroxy-PCBs 155
4-hydroxytamoxifen (OHT) 15–18, 72
hypospadias 41, 47, 54
hypothalamic pulse generator 87

I
ICI 164,384 72
ICI 182,780 12, 15, 17–19, 72
IGF, see insulin-like growth factor
imprinting 87
infant formula 101, 113
inhibin 146
insulin-like growth factor (IGF) 3, 11–12, 18–19
interactions
–, additive 43, 49–53
–, synergistic 49, 53
in utero exposure
–, to BPA 145–146, 148
–, to DES 132, 146
–, fetal positioning effects 131, 145–146
ipriflavone 106
–, biological effects 122–124
–, chemical structure 107
iprodione 57
isoflavones 14–15, 101
–, anticarcinogenic activity 118
–, chemical structures 67, 104
–, exposure 111–113
–, occurrence and metabolism 104–106
isolariciresinol 68

J
Japanese medaka 139

K
kaempferol 69
kelthane 57
Kepone, see chlordecone
ketoconazol 79

L
lariciresinol 68
LBD, see ligand binding domain
Leydig cells 146
ligand binding domain (LBD) 4-5, 9, 11, 16
lignans 14-15, 101
-, anticarcinogenic activity 118
-, chemical structures 68, 103
-, exposure 111-113
-, occurrence and metabolism 103-106
linuron 79
LNCaP 33
low dose effects 145-146, 148-149
luteal phase 90

M
male genital tract 131
-, epididymis 45, 139, 146
-, feminization 132
-, Leydig cells 146
-, preputial gland 146
-, prostate 45, 47, 54, 132, 139, 145, 149
-, seminal vesicle 45, 47, 49, 139, 147
-, Sertoli cells 146
-, sperm count 47, 54, 131-132, 139, 146, 156, 164
-, testis 131, 139, 146
mammalian androgens 71
mammalian-based reporter gene assay 31-33
mammalian estrogens 65
mammalian lignans 68
mammalian progestins 71
mammary glands 131, 146-147, 149
MAPK, see mitogen-activated protein kinase
matairesinol 68
MCF-7 cells 33-34, 48, 187, 189, 158
MDA-MB-231 cells 158
medroxyprogesterone acetate (MPA)
-, chemical structure 44, 74
-, inhibition of AR N/C interaction 56-57
megestrol acetate 74
melengestrol acetate
-, chemical structure 74
-, as a growth promotor 73
memory 95
menarche 87
menopausal hot flushes 94
methoxychlor 41, 46-47
-, chemical structure 44, 77
-, inhibition of AR N/C interaction 57
-, metabolite 44, 48
4'-methylcoumestrol 68
methyltestosterone 73
methyltrienolone (R1881) 44, 47, 54, 56
mibolerone 56
mifipristone (RU 486) 74
mitogen-activated protein kinase (MAPK) 11, 18
mixtures 43, 49-53
mosquitofish 57
MPA, see medroxyprogesterone acetate
multigenerational studies 58, 147
multiple chemical exposures 49
mycoestrogens 69-70, 107-111

N
naringenin 69
N/C interaction assay 55
non-additive interactions 159
nonoxynol 134, 138, 149
nonylphenol 57, 132-139
nonylphenol polyethoxylate 133
nordihydroguaiaretic acid 68
norethisterone 74

O
octylphenol 57
octylphenol polyethoxylate 133
Office of Economic Cooperation and Development (OECD) 41, 58
OHT, see 4-hydroxytamoxifen
organochlorines 14
ovarian follicle 88
ovariectomy 90
ovary 131, 139
ovulation 99
oxandrolone 56

P
paper mills 57
PCB, see polychlorinated biphenyls
PCDD, see polychlorinated dibenzo-p-dioxins
PCDF, see polychlorinated dibenzofurans
2, 2', 3', 4', 6'-pentachloro-4-biphenylol 160
peroxidase activity 158
pesticides 1, 14, 46, 132-134, 155

phage display assay 34
phenol red
–, chemical structure 75
–, estrogenicity 75–76
–, impurity 75–76
phthalates 14
–, activity in yeast-based assay 187–188, 193
–, chemical structures 78, 172–173, 190
–, effects on reproductive development 169, 183, 197
–, effects on vitellogenin 187, 191–192, 197
–, industrial consumption 171
–, in food packaging 177
–, metabolites 173, 178–183, 188, 193–195
–, monoesters 179–182, 184, 188, 194, 196
–, multigenerational studies 184
–, receptor binding assay 187–189, 195
–, solubility 171, 173, 175, 181, 185–186
–, structure-activity relationship 189
–, testicular toxicity 182
phosphorylation 8
photo-oxidation 144
phytoestrogens 1, 9, 12, 15, 18, 67–69, 101, 103–107
piceatannol 69
piceid 69
pinoresinol 68
placenta 144
plasticisers 171, 173
polycarbonate plastics 139, 141, 143
polychlorinated biphenyls (PCBs) 15, 77, 156, 169
–, chemical structure 77
–, hydroxylated metabolites 157
polychlorinated dibenzo-p-dioxins (PCDDs) 77, 156
polychlorinated dibenzofurans (PCDFs) 77, 156
polyvinylchloride (PVC) 173, 176–177, 179, 182
postmenopause 94
pratensein 67
predators 149
premature thelarche 87
premenopause 93
8-prenylnaringenin 69
preputial gland 146
preputial separation 58
procymidone 47
–, as antiandrogen 52
–, chemical structure 44, 79
–, inhibition of AR N/C interaction 57
progesterone
–, chemical structures 71
–, receptor 132, 139, 142, 158
prolactin 49, 140, 148
prostate 45, 47, 54, 132, 139, 145, 149
prunetin 67
psoralidin 68
PVC, see polyvinylchloride

R
rainbow trout 137, 138
raloxifene 17, 72
relative binding affinity 29, 160
resveratrol 106
–, chemical structure 69, 107
–, health effects 119–121, 124
rivers 132, 135–137, 144
RU 486, see mifipristone

S
Safe Drinking Water Act 41
salmon 139
secoisolariciresinol 68
secoisolariciresinol diglycoside
–, as antioxidant 120–121
–, biological effects 115–118
–, chemical structure 103
–, metabolism 105–106
selective estrogen receptor modulator (SERM) 12, 14–20
seminal vesicle 45, 47, 49, 139, 148
SERM, see selective estrogen receptor modulator
Sertoli cells 146
serum 157
sewage 132–133, 135–136, 138
sex hormone binding globulin (SHBG) 84, 140
sex reversal 157
sexual differentiation 43, 45, 48, 140, 146
SHBG, see sex hormone binding globulin
SHR, see steroid hormone receptor
skeletal ossification 145
skeletal system 94
skin atrophy 95
β-sitosterol 57
soybeans 104–105, 113, 118, 120
sperm count 47, 54, 131–132, 139, 146, 156, 164
spermicides 134, 148
SRC-1 10–11, 17
steroid hormone receptor (SHR) 3, 4, 10–12
steroidal estrogens, see mammalian estrogens

steroid receptor RNA activator (SRA) 10
structure-activity relationships 159
sulfatase deficiency 86, 93
synergistic effects 49, 53

T

taleranol, see β-zearalanol
tamoxifen 12, 14, 16, 18–20, 72
TeBG, see testosterone binding globulin
testicular atrophy 90
testis 131, 139, 146
testosterone 43–45, 56, 71
testosterone binding globulin (TeBG) 7
2′,3′,4′,5′-tetrachloro-4-biphenylol (HO-PCB4) 157–159
3,3′,4,4′-tetrachlorobiphenyl 163
3,3′,4′,5-tetrachloro-4-biphenylol 163
2,3,7,8-tetrachlorodibenzo-p-dioxin (TCDD) 77
tier I testing 41, 58
tier II testing 41
thoracic nipples 48
thyroid
–, function 163
–, hormone 42
2′,4′,5′-trichloro-4-biphenylol (HO-PCB3) 157-159
toxaphene 164
toxic equivalency factor 49, 51
transthyretin 163
17β-trenbolone 73
tyrosine kinase 120

U

uterine defense mechanisms 89
uterotrophic assay 147
uterus 146–147, 149

V

vagina 132, 146–147, 149
vas deferens 45
vinclozolin 41, 46–47
–, as antiandrogen 52
–, chemical structure 44, 79
–, inhibition of AR N/C interaction 57
–, metabolites 44, 52
vitellogenin 135, 138–140

W

waste water treatment 134–137, 143–144
water
–, contamination 141, 143
–, drinking water wells 134–135, 143
–, groundwater 137
–, rivers 132, 135–137, 144
–, waste water treatment 134–137, 143–144
wildlife 132, 148–149
–, alligators, Lake Apopka 41, 132
–, predators 149
witch's milk 86
Wolffian ducts 45
wool mills 135, 138

X

xenoestrogens 20, 163
Xenopus laevis 140, 146

Y

yeast-based steroid receptor assay 30–31

Z

α-zearalanol 70
β-zearalanol 70
zearalanone 70
α-zearalenol 70
β-zearalenol 70
zearalenone 70
zeranol, see α-zearalanol
ZR-75 cells 187–188

MIX
Papier aus verantwortungsvollen Quellen
Paper from responsible sources
FSC® C105338

FSC
www.fsc.org

If you have any concerns about our products,
you can contact us on
ProductSafety@springernature.com

In case Publisher is established outside the EU,
the EU authorized representative is:
Springer Nature Customer Service Center GmbH
Europaplatz 3, 69115 Heidelberg, Germany

Printed by Libri Plureos GmbH
in Hamburg, Germany